WHAT
IS THE
GRAND DESIGN?

PETER
BRIGGS

WHAT
IS THE
GRAND DESIGN?

DAVID McKAY COMPANY, INC.
New York

The author gratefully acknowledges the permission to reprint from
the following:
From the Brunn Memorial Lecture given by Dr. Arthur E. Maxwell
of the Woods Hole Oceanographic Institution at the Sixth Session
of the Intergovernmental Oceanographic Commission, UNESCO,
Paris, 1969. Permission to reprint granted by Dr. Arthur E. Maxwell.
From *The Cloud Forest* by Peter Matthiessen. Copyright © 1961 by
Peter Matthiessen. All rights reserved. Reprinted by permission of
The Viking Press, Inc.
From *Life Around Us* by Fritz-Martin Engel, English translation
copyright © 1965 by George C. Harrap Limited and Thomas Y.
Crowell Company. All rights reserved. Reprinted by permission of
Thomas Y. Crowell Company.
From *My Philosophical Development* by Bertrand Russell, pub-
lished in 1959 by Simon & Schuster, Inc. All rights reserved. Re-
printed by permission of the publisher.
From *The Life of Insects* by V. E. Wigglesworth. All rights reserved.
Permission granted to reprint by the publisher, Universe Books.
From *The Theory of Evolution* by John Maynard Smith © 1968.
Permission to reprint granted by the publisher, Penguin Books Ltd.

LIBRARY OF CONGRESS CATALOG CARD NUMBER: 72-75464

MANUFACTURED IN THE UNITED STATES OF AMERICA

WHAT
IS THE
GRAND DESIGN?

The question "Is there a grand design?" may seem outrageous. It asks, can a purpose be found for the incredible unfolding of evolution, the vast profusion of forms of life? (What was the purpose of the dinosaur?) Is there a meaning in the marvelous story of how living things came to be? Or can we find some significance in the history of the restless earth itself. The outrageous answer here is "Yes."

The wonder of how man came to be, three billion years after life sparked on earth, can never be known completely. The great forces shaping the planet have erased many of the clues to the past. Still, new bits of evidence constantly turn up and the narrative is never finished. Yet now we understand a great deal concerning our beginnings and it is a fascinating tale.

But can we, by unearthing how we physically arrived, give a reply to the really fundamental question? Can the laws shown to exist for evolution demonstrate a purpose for man? The reply is in the affirmative. No matter how blind man's ascent seems through all the convulsions, extinctions, and disasters, as a result of his success there now exists a creature with power to give future evolution a direction.

1

I would find it very exciting to go far back in time, briefly, and see the people responsible for my being here. My ancestors. To watch the first man carrying my name in America come ashore at Plymouth, Massachusetts, with his wife and eleven children after the many weeks long voyage across the Atlantic on board the *Fortune.* What kind of a man was he? I would enjoy even more going back, not just three hundred years but five hundred, and observe, somewhere in Yorkshire the wool carders who gave me my name, the earlier men and women who finally caused my existence. Would I look like them? Would I even understand their speech? What did they consider the purpose of their lives? Even more would I like to return, for a short while, to the time when my ancestors in England and Scotland first met the invading Romans. Were they brave or did they cohabit and collaborate? Did they die resisting and, if so, why? Perhaps my ancestors had not even reached Britain at the time of the Romans but instead were digging clams in Denmark or fishing in the cold water off Norway. Who *were* the people I came from? I can never know very much about them but that is why history fascinates me.

An uncle of mine wrote a book once about a progenitor, the Black Douglas, a Scottish lord and warrior who successfully battled the English but died in Spain at the hands of the Moors in 1330 while carrying the heart of his King, Robert I, to bury in Palestine. So I know about him but I am not interested in my remote past just in order to find noble families that might suggest a fake glamor for me. I would like to know all about the people I came from because, if they were all real to me, that would somehow give me a firmer foundation, a feeling of continuity, a reason for being. If I could know that these people acted with a purpose, I would feel less as if I am here on earth purely by accident.

Then, too, there is the simple matter of curiosity. Vast amounts of scientific labor have been expended in reconstructing the story of the world and the life on it from its very beginning. The people who have studied the past so carefully have done a marvelous job, but I want to know more. Of course, the millions of creatures in the long span of evolution, whose existences resulted in my being, had no thought except to stay alive and reproduce but my curious eyes would like to see them alive in their strange worlds anyhow. Perhaps the scientists have not figured it all out correctly. Actually, I believe that they have but I would like to see a documentary film showing the first animal to make the move from the water to land. It would be a smash at the box office.

When one is told that the first creature to live out of water existed 350 million years ago, such information is emotionally meaningless. Our sense of time is very limited. We have a feeling for minutes, months, even decades and whole lifetimes but what sense can we make of a thousand years, much less whole geological ages. Until quite recently most people believed that the world began when the Bible said it did. As calculated by Archbishop Ussher of Ireland, and accidentally printed in early editions of the St. James Version, the world was created at nine in the morning,

October 26, in the year 4004 B.C. This was a very long time ago but still possible to comprehend.

But starting with Leonardo da Vinci, who kept his ideas secret in his notebooks, and an Italian named Nicholas Steno who began publishing his ideas in 1669, thinking men in Western Europe began having new ideas about time. These ideas were often dangerous during a period when the dominant Christian theories about the earth and its beginning were enforced by the rough hands of the law.

During the eighteenth and early nineteenth centuries, particularly in France and England, men like Baron Georges Cuvier and William Smith began looking at the rocks around them and wondering how they got there. They studied the solidified remains of marine and land animals that were imbedded in them. These rocks were obviously the final resting place of all sorts of once-living creatures. Often they resembled animals now on the face of the earth but more frequently they were quite different. Sometime, obviously, these rocks with their remains of sea creatures had been under water although they were discovered on what is dry land now, far away from the seashore. Some of these early geologists and paleontologists began to see that the creation of the rocks with their remains from the past was quite a long process and these early scientists began to demand much more time for the history of the natural world than the Bible had allotted to it.

Since the days of Dr. James Hutton, who might be said to have founded geological science when he brought out his *Theory of the Earth* in 1785, the date for the beginning of our planet has been steadily pushed backward. In 1941 a textbook set the date at about two billion years. Very recently, by using the new method of radioactive dating, creation has been placed at 4.5 or 4.6 billion years ago. Rocks brought back from the moon and studied by this method give almost exactly the same date.

How can our finite minds comprehend such lengths of time? And does it really matter anyhow? Is it important to know when the world began? Or what the sun is, how the ocean got there, why animals exist, and why there are so many different kinds; why we human beings ourselves are born and live our lives here? Primitive men thought so and every ancient religion, every racial folklore, has its own story of creation, its own try at explaining the enormous facts of nature around us. As in the Bible, the story of creation is the beginning of the whole religious story, the foundation of the entire moral code. It was the Hebrew-Christian God who, after making the Earth, gave man a creed to live by.

But since the Biblical version of natural events has been found to be less than accurate, where is the basis for Christian morality? Can a foundation be discovered by looking at the evidence found in nature itself? Might the study of rocks, of animals living and fossilized, by the method known as "science," give us a sure base for a moral code? *Can morality be found in a grand design that can be discovered in the vast history of the earth?* Many human beings have hoped so and have devoted their lives to discovering the natural laws of the world around us. As scientists they are obliged to consider nothing but the real evidence they can find about them, so they must deal in particulars before they dare make generalities. This is congenial because there are many fascinating questions smaller than the existence of a grand design itself and finding the answers to these smaller riddles is felt to be a rewarding occupation to many people. These are the fortunate people who are curious. Like little children, they want to know what is behind the mirror.

Humans share the quality of curiosity with cats, crows, monkeys, and other insatiable creatures. The Elephant's Child in Kipling's *Just So Stories*, "a new Elephant, was full of 'satiable curiosity and that means he asked ever so many questions." He filled all Africa with his 'satiable curiosity, according to Kipling. One fine morning the Elephant Child asked a new fine question he had never asked before. "What does the Crocodile have for dinner?" Everyone he asked told him to "Hush" until finally he asked the Crocodile himself. The Crocodile answered by saying "I think I will begin with the Elephant Child himself" and clamped his jaws on the end of his button nose to pull him into the water. He pulled and pulled and the Elephant Child pulled too until the Elephant's nose was stretched out of shape. At last, the Elephant Child freed himself with the aid of the Python-Rock-Snake. But his nose never did shrink back and that is How the Elephant Got His Trunk.

That Kipling wrote such stories, to give entertaining explanations of nature to children, and that the stories have been such a success, makes it seem that there is a genuine curiosity here. Kipling covered only a few of the strange animals with his tales. There are many, many others, perhaps even stranger animals than elephants, animals interesting both because one may be curious about how they got that way and interesting because an explanation of their development is in part an explanation of evolution itself.

Scientists sternly warn against anthropomorphism, the habit of thinking that animals have just the same motives as human beings, but it is difficult not to be anthropomorphic about the amazing bower bird. This resident of New Guinea and Australia is a close relative of the gorgeous birds of paradise but all bower birds are very drab. Like the bird of paradise and the scarlet Cock of the Rock in British Guiana, the bower is an arena bird. The males live apart from the females, in clans, and each bird has his own breeding station.

Though allied with other species, a 19th-century naturalist is said to have decided that all birds should be divided into two categories; bower birds and other birds. Male bower birds erect large walled bowers made of sticks and decorate them with bright objects. One species even makes paint by mixing saliva with fruit pulp and, using a brush it produces from bark fibers, paints the inside of the bower. (A rare example of an animal using a tool.) Some species create towers as much as nine feet tall, with tepee roofs and rooms inside. In front there is a courtyard kept scrupulously clean and beautified with bright stones, shells, colored berries, and fresh flowers that are renewed every day. There may be half a dozen such bird palaces, all within sight or at least sound of each other, and each decorated a little differently according to the bird's individual style. During the mating season each male tends his bower carefully and waits for the female to inspect it.

When the female arrives, some males crouch on the ground as if tending a non-existent nest. In other species the male suddenly begins to act like a baby bird, fawning and begging for food. The males then often court the females by bringing a berry as a gift. If the female gives the suitor a peck on the head, he knows he is accepted. Mating begins and is over within a few seconds. The females then depart. She flies high into a tree and begins to build another nest, a simple practical one in which to lay eggs and care for the chicks when they are hatched. The male has nothing more to do with the matter.

Arena behavior occurs in a number of bird species that are only distantly related and even live on different continents. The ruff, a sandpiper of northern Europe and Asia, is an example of species where the males and females live apart the entire year except for a few minutes during mating season. Each male has established his own territory and it is up to the female to make a choice. (The males do not fight at this time. They have already done so settling their territorial rights.) The splendid bird of paradise keeps his arena absolutely clear of forest leaves, does not build, but does chew off leaves above his ground so that shafts of sunlight can get through to shine on his brilliant feathers.

What use is the arena in terms of evolution? The explanation seems to be that the males of these species are so gaudy that they are an easy, conspicuous mark for predators. If they were to remain tied to a nest during all the egg-hatching and infant-feeding time, they would be too good a target. Natural selection favors a species in which brightly colored males are freed from nesting duties. Sexual selection can operate more freely and the males develop ever more gaudy plumage to attract females.

The bower birds, however, have advanced even more in this evolutionary direction. The males, almost as inconspicuous as the females, have transferred their display to their bowers, and after mating can leave their showiness behind and retire inconspicuously into the forest. Yet they have not yet abandoned all their nesting instinct and the bower is that nest, a survival of bird behavior that has gone on for probably millions of years. Bower birds have replaced their glittering but dangerous plumage with natural jewelry which the females have learned to like just as well. It may not be too far-fetched to call the behavior of bower birds an example of what is known technically as "convergence," the parallel development of the same character in two species that are not genetically related. In this case, of course, the bower bird's display seems very like the social evolution in human beings in which individuals decorate their bodies in all sorts of startling ways in order to be sexually attractive. Sea shells from far distant waters, undoubtedly used for ornaments, are often found in Neanderthal caves.

The case of the male stickleback is a case of the very contrary ways that animals have evolved in relation to matrimony. Sticklebacks are fresh-water fish. In general, among fish, the only role the male has in reproduction is to spray his sperm over the eggs after the female has laid them. This scrappy little male, however, dominates the entire breeding procedure. In the spring he strikes out alone from the school of sticklebacks with whom he otherwise associates. He selects a territory which he briskly defends against any invader, male or female. At some signal given him by his hormones, he digs a shallow depression in the sand and builds a nest over this made from stringy algae, cemented together by a secretion from his kidney. Then he builds a hole in the nest by swimming through it. The nest is smaller than the fish so that both the head and tail stick out from either end.

The stickleback is normally a gray color but at a certain time he turns bright red. At this time he is most aggressive and possessive about his territory. (Sticklebacks in an aquarium in London, who could see the street outside through the glass aquarium walls and the windows, made violent attacking motions when they saw a bright red Royal Post Office truck parked there. They thought it was a red male stickleback invading.)

The female of the species meantime has been gestating some fifty to hundred eggs and instinctively, at the time to spawn, she approaches the red male. He alternately attacks her, and then controls himself, and finally leads her into the nest he has built. Once she is inside he prods her tail rhythmically with his head and this motion causes her to lay the eggs. Now the male drives her out of the nest, enters it himself and deposits his sperm. Then, if the female is still nearby, he chases her away and waits for another gravid female. He repeats the procedure with her and will continue, with four or five females, until his nest is full of eggs.

Then his red color subsides while, in the days needed for the eggs to hatch, he hovers over them, fanning the water with his fins so that it will brush over the eggs, keeping them clean and supplying them with fresh oxygen. When the little fish are hatched he herds them together for a day or so, bringing back any strays in his mouth. Then finally the young join others of their generation and the adult male returns to the school of sticklebacks.

This fish has been most extensively studied by Dr. Nika

Tinbergen of Oxford who is particularly interested by the hostility the male displays during courtship. This is purely innate behavior and one that is displayed by several kinds of birds. Dr. Tinbergen finds similarities in this hostility with that which is sometimes found in mating human beings.

The story of the American beaver illustrates a number of points in the laws of evolution. Beaver history also tells a good deal about how a regional ecology changes and of the role of certain animals in the evolution of human society. Man owes great debts, which they have mostly forgotten, to many "wild" animals.

Although one may not want to use such an ugly word for such an admirable creature, the taxonomists who classify species say that the beaver is a rodent. Like his rat cousins, he must gnaw very frequently to keep his teeth from growing too long. So relentless is the growth of their teeth that if, through accident or deformity, one tooth fails to meet its mate and thus be worn down, it may keep growing outward, curve back into the mouth, prevent the jaws from closing and finally cause death by starvation.

The ancestral beaver was living in North America at least fifty million years ago; the oldest rodent fossil, the only one of Paleocene days, was found at Bear Creek, Montana. Before this the region had a thriving species of now fossil animals called multituberculates. Their way of making a living was almost the same as that of the rodent family and a common rule of evolution is that no two species can long inhabit the same area if their habits are too much alike. When the rodents appeared, their rivals had to go because the rodents were much more efficient.

American beavers (there is a close European relative) about a million years ago began to illustrate something known as Cope's Rule. This says that the body size of a species will continue to increase as evolution progresses. In the case of the beaver this rule was reenforced by Bergmann's Rule, that the body size increases as the climate becomes colder. Arctic animals are usually larger than tropical ones to whom they are related, because a larger animal loses less of his body heat to the atmosphere than a smaller one. Sometime between the Pliocene and Pleistocene epochs, the time of Ice Ages, the ancestral beaver evolved into the size of a bear, about 700 or 800 pounds.

The modern beaver, whose luxurious fur had much to do with the advance of Europeans into North America, is an amazingly industrious creature. Many mammals are nest builders but no other builds anything so imposing as a beaver dam. To build his home the beaver begins by gnawing down willow or alder branches, carrying them to a strategic place in a stream, and fixing them in place with his paws. Then he adds mud, gravel, and stone, then another layer of brush and saplings, then more mud and stones, until his construction is high above water level. The beaver prefers young trees at the water's edge but will, if he must, gnaw down trees as much as ten feet in circumference and drag them to the water. Sometimes beavers even dig canals to float the logs. One such canal was 750 feet long. The beaver dam, of course, is a nest for the newborn and adolescent animals and a storage place for fresh greenery that keeps the animals alive as they hibernate during the winter. Toward the end of winter, the beavers survive on wood bark.

The unusual tail of the beaver is used as a paddle for swimming, as a prop to sit back on while gnawing a tree, and as an alarm signal. When a beaver slaps the water with his tail, making a very audible "plomp," every other beaver in the pond dives to safety. (The beaver's house is so strong that predators such as wildcats cannot tear their way in.)

The larger the animal, the longer its gestation period and as the largest rodent, the female beaver carries her young for three months. Young beavers live in and near the dam for two years, learning the complicated arts of beaver living.

The animals have developed far beyond the point where all their behavior is instinctive. They must learn.

In the spring of their second year the young beavers are driven from home. They must go out into the world, find their own mate and their own place on a stream to build a dam. This serves the function of radiating the species and keeping the immediate population down to the point where there are enough trees to feed all the animals present. Some young beavers are reluctant to leave the security of their dam and the otherwise-model parents have been known to kill offspring who refuse to leave. (This kind of killing, to control the population, is known in a number of mammal species.)

Ecologically, the beaver dam and the pond it creates are a boon to many other creatures. Algae develops, food for birds and fish, insects breed on it, other mammals will use it for a water hole. In time the pond begins to eutrophy, grow thick with silt and weeds. The beavers abandon their dam and its now overgrown pond. In time the pond dries out, the land where it lay now being good, rich soil ideal for farming. Many prosperous American farms were established on land made ready for them by beaver dams.

The fossil evidence is too poor to show just how the beavers developed their wonderful way of life. Like all their family, they inherited the traits of gnawing and nest-building but beavers went far beyond the rest of their family in evolving a complex style. One cannot tell whether they first appeared in Europe or North America but certainly they flourished in the New World. Finding a congenial source of food in leaves and young trees, they settled along the stream beds. Needing a nest they used these same trees as the most obvious building materials. Placing the sturdy nest in a stream gave added security and a storage place. Their teeth became specialized to suit their diet. All rodents can swim but beavers perfected this ability and those whose tails were a little bigger could use them a little better in the water and so, through the long process of natural selection, broader tails made their owners more successful in the water environment. And here is a basic principle of evolution. A species adapts a particular feature and specializes in it for the simple reason that the feature is successful. If a door is locked, only those who learn to use a key are able to open it.

The beavers' development of a soft, rich coat to protect them from the cold water and the long winters almost led to their ultimate downfall. The first Europeans in America brought home these wonderful furs and the demand for them was intense. King Charles II of England in 1638 made the use of beaver fur compulsory in the manufacture of hats. The French explored the St. Lawrence River, the Great Lakes, and finally the Rockies in search of beaver fur. The cities of Quebec and Montreal began as trading posts. The Hudson's Bay Company of Canada was formed to trade in beaver. A Company sale in November 1743 disposed of 36,750 beaver pelts and in the same year 127,080 beavers pelts were received at the French port of La Rochelle. A steel trap was invented in 1823 to capture beaver and the rate of annihilation increased. The great fortune of the Astor family was begun in the fur trade of John Jacob Astor who shipped pelts from Astoria in Oregon to China where the nobility paid enormous prices for them. Astor's flagship was, quite reasonably, named the *Beaver*. Surely the estimable beaver did not evolve so that some day he could enrich the invaders of North America, but exist the beaver did and nature's most greedy predator went after him. Having had such an immense influence on history, it seemed that the beaver would become extinct toward the end of the 19th century but, in one of the early conservation laws, he was declared protected. Beavers as a species have a strong will to survive (though such a thing is hard to prove in scientific experiments) and today they seem to be in no immediate peril.

In the case of the beaver, and among many other species,

man has become a major force in evolution because of his well-proven ability to cause extinction and to change the balance of an area's ecology as well. Man has also become an important factor in the whole web of life by manipulating evolution through control of the breeding of plants and animals. Considering the length of geological time they have been doing these things, human beings have been very efficient biological agents. However, there are today about 350,000 species of plants and perhaps 1,120,000 species of animals and thus far many of them have escaped his activities.

No plant or animal species, nevertheless, with the possible exception of some primitive bacteria and blue-green algae, has remained unaltered by all of the numerous forces that make species constantly evolve. A few, such as cockroaches, horseshoe crabs, the American opossum and perhaps sharks, are so successful that they have hardly changed at all in many millions of years. For most species, a very basic law is *adapt or perish*.

The root of this law is that the world itself is a very active place, constantly becoming something different. Glaciers cover vast portions of the earth and then recede. Mountains rise to great heights and at last are worn down to the roots by erosion. Shallow seas invade the continents. (The center of North America has been covered with salty water many, many different times.) Islands become isolated from the mainland, marooning the things living there at the time of separation. Continents move about so that at one time New York was on the Equator, and corals grew on the beaches of Alaska. Continents that once touched later moved apart and became islands on a cosmic scale. Land bridges that once existed have disappeared. These great changes in geology, so far, outstrip the puny efforts of man to control every affair.

It is hard to know which evolutionary forces were most important in shaping the Pteranodon, an extinct flying reptile with a body the size of a goose and a wing span of nearly 30 feet. (One may easily guess why this ungainly thing is extinct, however.) It is easy to see, now that Darwin has explained it, why the finches on various Galapagos islands have different kinds of beaks. They were isolated. One can fairly well understand how one single, pregnant fruitfly that landed on the new Hawaiian island of Maui 700,000 years ago had descendants that evolved into 200 distinct species. The fruitflies occupied a niche in the Maui ecology that no other insect already inhabited. (Just as a population explosion occurred among rabbits when they were introduced into Australia. There were no competitors.) A reason can be given for the existence of a type of fresh-water shark in large Lake Nicaragua in Central America. They were trapped when the lake, then an arm of the ocean, was suddenly cut off from the sea by a convulsive movement of land. The lake water slowly changed from salt to fresh and the sharks had time to adapt. An extreme case of response to changes in both weather and geology are the penguins of Antarctica. It is almost impossible to believe that these primitive birds actually migrated to such an inhospitable place. It is possible to accept the idea that they lived on the Antarctic continent when it was in a more northern, warmer latitude and that, being flightless, they earned their livelihood by fishing. (An earlier adaptation, of course, was that they gave up their wings for flying and used them for very fast swimming instead.) As the Antarctic broke off from Gondwanaland and drifted south, the helpless penguins had to follow their breeding grounds, to which they were attached by iron instinct. As the atmosphere became colder and colder, the birds adapted better and better insulation and learned to lay their eggs on land during the one or two months when this was possible. But one species, the Emperor penguins, went even further and learned to lay their eggs when it was obviously impossible; the long black night of winter when temperatures might be 70 degrees below zero. The Emperors might lose some eggs to the weather, but they were absolutely safe from predators!

(Some ancient amphibians and reptiles resident in the Antarctic, having no internal heat regulator like that possessed by birds and mammals, were unable to adapt to the increasing cold. Such species died out and their fossil bones only recently were discovered in the Trans-Antarctic Mountains.)

However, animal stories do not illustrate first principles when told at random and in the most general context. To see how such awesome productions as golden eagles, blue whales, polar bears, Bengal tigers, and Homo sapiens came to be, one must be orthodox and begin with what is known about the origin of life.

In its January, 1871, issue the magazine Scientific American published the following story. "Professor Huxley, the eminent naturalist, has sometimes been accused of exhibiting a pugnacious and acrimonious spirit. At one of the meetings of the British Association, Samuel Wilberforce, Bishop of Oxford, blandly asked him in the presence of a large audience: 'Is the learned gentlemen really willing to have it go forth to the world that he believes himself to be descended from a monkey?'

"Professor Huxley rose and replied in his quiet manner: 'It seems to me that the learned bishop hardly appreciates our position and duty as men of science. We are not here to inquire what we would prefer, but what is true.

" 'The progress of science from the beginning has been a conflict with old prejudices. The true origin of man is not a question of likes or dislikes, to be settled by consulting the feelings, but it is a question of evidence, to be settled by strict scientific investigation. But, as the learned bishop is curious to know my state of feeling on the subject, I have no hesitation in saying that, were it a matter of choice with me (which clearly it is not) whether I should be descended from a monkey, or from a bishop of the English church, who can put his brains to no better use than to ridicule science and misrepresent its cultivators, I would certainly choose the monkey!' "

2

*Philosophers enjoy the co-*nundrum "What is life?" Instructors in biology also seem to like dwelling on the problem, teasing their students by asking them why fire is not a form of life. It consumes, it grows, it reproduces and so on. Why isn't fire a living thing? Perhaps no satisfactory definition of life exists but in all common sense we know what is living and what is not. (Except in the case of viruses, however.) For practical purposes we know full well that life exists. "Why" it does is not an easy question for many people to answer. It may be a little simpler to know "how" life came to be on this planet.

The Bible disposes of the matter in several brief sentences but to some minds that explanation is not satisfactory. Even in the Middle Ages further ideas seemed necessary. To explain lower forms of life, the idea of "spontaneous generation" was conceived, the thought that there was some kind of miraculous change from non-living to living matter. A Dutchman in the 17th century wrote that mice are created by filling a jar with wheat and stuffing the top with a woman's dirty shirt. Father Kircher, a professor of science in Rome at about the same time, liked the idea of spontaneous generation for lower breeds because it settled a problem having to do with Noah and the Flood. "It is obviously pointless to give these latter forms a place in the *already encumbered Ark."* Obviously Father Kircher had some idea of the vast number of species in the world and wondered how one ship could handle them all, two by two. Then spontaneous generation disappeared as a theory when Pasteur showed in his microscope that these latter forms, such as bacteria, generated, not miraculously but simply by dividing.

A first date for the origin of life may have been established by the development of the geologic time scale, one of the great feats of the human mind. At the moment this scale proposes that the age of the earth is 4.5 billion years. At that time, almost everyone agrees, there was no free oxygen in the air. At present, only some bacteria and algae can exist without oxygen. Life as we know it could not have been possible at first.

Fifty years ago a popular theory of the origin of life began with the idea that the first atmosphere was made of ammonia and methane gas. This, presumably, was because that is the present atmosphere of Venus. Some students of old rocks say that the evidence contradicts such an atmosphere. However, most scientists agree that the first atmosphere probably had organic compounds in it, amino acids. The compounds were put together by inorganic means. What caused them suddenly to spark to life? God, if you like, or then something else still to be explained.

A meteorite that fell to earth not long ago, on examination, turned out to have the same age as the earth. It contained the kind of compounds visualized as being in the earth's early atmosphere. Yet the meteorite came from outer space, perhaps from outside the solar system. Maybe life did not originate on earth at all!

The oldest known group of sediments in the world were found in what has been called the Fig Tree Formation, between Swaziland and South Africa. In 1965 Elso S. Barghoorn of Harvard took samples of chert (a flint-like rock

18

with organic associations) from the Fig Tree for study under an electron microscope. He found two extremely primitive forms of life in the chert, one resembling a blue-green algae, the other looking like bacteria. Tests show that the Fig Tree Formation has an age of 3.2 billion years and so these earliest living things are a billion years younger than the earth itself. It took a very long time apparently for life to originate.

These earliest organisms could not exist without free oxygen and so the great revolution must have already occurred, the change without which any higher form of life would be impossible. This revolution was the development of photosynthesis, the process by which plants derive their energy from the sun. The bacteria probably did not release any free oxygen in the atmosphere but the algae almost certainly did. These first organisms surely formed in water deep enough to shield them from the sun's most damaging rays. As more and more free oxygen was released into the atmosphere, however, the new kind of atmosphere with oxygen in it would screen out the worst effects of the sun. Plants could venture forth a bit more bravely.

The oldest fossils so far discovered in North America are about 2 billion years old, They are imbedded in what has been named the Gunflint Formation near the northern shore of Lake Superior in Ontario. These Gunflint rocks have yielded 12 different species of primitive plants, plants that by this time had developed chromosomes, the basis for sexual reproduction and thus for a rapidly increased rate of evolution. (No organism in the world is exactly alike any other so when the two organisms mate the result will be organisms that are not an exact duplicate of either parent. This creation of diversity is the basic function of sex, as far as evolution is concerned.)

Similar, unbelievably small, evidences of the dawn of life have been found in Australia but such relics of more than two billion years of life are extremely rare. The record of the rocks in this long period called the Pre-Cambrian is a good deal more instructive.

This Pre-Cambrian landscape would have been almost completely unrecognizable, as the world he knows, to any present-day human being. (Of course no human could have viewed it since there was not enough oxygen in the air for him to breathe.) No green thing could be seen on any of the bare rocks. At the time of the African Fig Tree Formation three billion years ago, North America consisted primarily of Ontario and a very small portion of northern Minnesota and North Dakota. There was considerable outpouring of lava through the thin crust of the earth (as yet not heavily covered with sedimentary and metamorphosed rock). There were also two Ice Ages in this part of North America, now called the Laurentian shield. There was also a mountain range, that may have been as grand as the Rockies, which lay across what is now the border between Canada and the United States. It was worn down to its roots long before the period was over. During the Pre-Cambrian era the continent continued to grow by accretion but it by no means reached its present borders, particularly on the Pacific side or in the southern states.

In the rest of the world the only Pre-Cambrian rocks exposed today are in eastern Africa, the Baltic region of Europe, a rather small area in Siberia, the eastern portion of Brazil, and the western half of Australia. (The rocks in Pre-Cambrian Swaziland are very similar in chemistry to the rocks now being issued along the mid-ocean ridges.) Though many Pre-Cambrian rocks are doubtless buried under later accumulations, it is likely that all the continents were smaller then. It is also very likely that all of them were massed together in one great mega-continent.

Some studies have shown that during part of this long period the North American continent was much closer to what is now the North Pole. At other times part of the continent was below the Equator and tipped so that the East coast was face down; that is, what is now the western edge of the continent was then the northern edge and the eastern side was then the southern bottom. And, for a time during the Cambrian era, the European Equator ran from Norway across Asia to the China Sea. Spain was at 50° South latitude, the present position of Tierra del Fuego, the frigid southern tip of South America.

From the study of ancient magnetisms of the earth, it seems that at one period during the Pre-Cambrian era the North Pole lay in the middle of the Pacific Ocean. The most likely explanation for all this is not that the earth's axis of rotation has changed, but that the earth's solid crust has slipped slowly over the more fluid mantle beneath so that, at different times, different areas of the earth were brought into the polar position.

Yet the North Pole in former times does not seem, by the study of old magnetism, to have lain in the same direction from each continent. The explanation must be that, while all the earth's crust has been moving around, the crusts have also been moving in various different directions around *on it*, taking positions that were much different in relation to each other than the continents we have today. Such motions are hard to visualize. We have learned the present map of the world to be a *fact*. Apparently it is only a temporary fact and our solid earth only solid in the terms of a few human lifetimes.

A great philosopher has observed that we can command nature only by obeying her laws.

—Sir Charles Lyell, Principles of Geology, 1875.

3

We know that the bits of organic evidence proving life existed on earth 3 billion years ago are very rare; that they can be seen only through the world's most powerful microscopes. This remains true about every fossil in the long, long period known as the Pre-Cambrian that ended suddenly 600 million years ago. Then dramatically a number of living things developed hard parts that have been found in rocks all around the planet. The beginning of this Paleozoic, when fossils suddenly become quite abundant, has been compared to raising a curtain on the theater of life.

The comparison is only relative, however, since this opening scene occupied something like 10 or 20 million years; yet not much geologic time in comparison with the 300 million years that the Paleozoic era took to reach its climax. In the first subdivision of this time, called the Cambrian (of 100 million years' duration), almost all the basic categories of life made their appearance in primitive, though multi-cellular forms. They must have all been evolving for quite a while and then, practically all together, they changed and began to use a metal, calcium, as part of their system. Why calcium became a fundamental building block is not easy to understood since magnesium is three times more common in sea water. What caused the fundamental switch can only be guessed at. Did all the animals acquire armor in defense against each other, was it as protection against the sun's rays, or was it because a firm structure improved their ability to swim or crawl about and thus find new sources of food?

The shallow Cambrian seas contained protozoa such as algae, bacteria, and the tiny, shelled foraminifera whose changing shapes through all the ages are a key index for dating the sediment. These seas also had sponges, worms, brachiopods (some of which slightly resembled clams) and echinoderms, a class to which modern sea urchins and starfish belong. The dominant phylum was the Arthropoda, which later gave rise to the insects. In the Cambrian period the leading arthropods were the trilobites. These ancestors of today's horseshoe crab developed more than 2000 species during Cambrian times; certainly a most vital family and one whose evolution came about very rapidly. And this demonstrates one of the great generalizations about evolution. When a group takes on some way of life that has not been exploited, or moves into a virgin area, changes and adaptations come along quickly as the group takes every possible advantage of the new situation.

The first trilobites to develop a shell, very much like that of a lobster, must have had a decided advantage over less well-protected and coordinated cousins. (This shell is called —and it has a poetic sound—a "chitinous exoskeleton.") In defense against this new equipment other trilobites would also have to develop chitinous exoskeletons or lose out in the battle for survival.

The world which the trilobites took over so explosively hardly resembled the modern world at all. Fossils of the species found in Cambria in Wales (where the name of the period came from) so closely resemble trilobite fossils found in similar strata as far west as Arizona, that North America and Europe must have been much closer, or even joined, at that distant time. The Cambrian Equator ran from the northern tip of Greenland down past the western side of Hudson's Bay, crossed the present U.S. border in North Dakota, went down through Texas, across Mexico, and entered the Pacific at a point near the present-day town of Manzanillo.

North America was divided into three parts. In the east a high region that has been named Appalachia arose over what is now the Piedmont and Coastal Plains. No one is certain how far to the east of this it extended but the erosion of this mountainous rise poured great masses of detritus into the Appalachian geosyncline. (A geosyncline is a depression or trough which constantly fills with sediment and usually subsides under the weight.) This Appalachian geosyncline was in the position of the present mountain range and the materials forming this range came from the time when the region operated as a basin for deposits. On the west side of North America was another high, mountainous area called Cascadia, on the site of the present Coast Range and it, too, filled up a geosyncline, called the Cordilleran, that was later the source for the Rockies. Before the close of the Cambrian period, a third, similar set of features began to appear to the south. These were highlands geologists have called Llanoria. They covered southern Louisiana, Texas, and northern Mexico and stretched out into the present Gulf. The geosyncline north of this, several hundred million years later, produced the Ouachita (Wichita) mountains centering in Oklahoma. The geosynclines were generally under water and, as they depressed, vast, shallow inland seas developed.

Simultaneously, another great sea, perhaps as large as the Atlantic, existed beyond where the Mediterranean lies today as one of its last remnants. This sea, called Tethys after the mythical wife of Oceanus, may once have been open all the way from southern Europe to Indo-China and certainly covered the region that now contains the Himalayas. (Fossil sharks' teeth have been found in these mountains at 20,000 feet and the range must be formed out of a Tethys geosyncline.) The Tethys Sea was originally the concept of an Austrian geologist, Eduard Suess, who proposed that all the southern continents had once been a single area he called Gondwanaland, after a province in India. The Tethys Sea separated Gondwanaland from Laurasia in the north, a continent made up of Asia, Europe, and probably North America. The Tethys may have reached from Mexico to the Pacific. It lay in an uneasy shear zone between the Baltic and African shields that had existed since the Pre-Cambrian period. Today the action between the two shields, now often referred to as plates, has created the system of mountain ranges that run from the Pyrenees, the Alps, and the mountains of the Near East to the mountains of central Asia. In addition to the Mediterranean, the Black Sea and the Caspian Sea may also be relics of the once great Tethys.

If the living habits of their modern descendants can be taken as any indicator, the little creatures of the Cambrian era must have lived mostly in warm, almost tropical seas, even at what today are quite northern latitudes. As the period moved on, the mountains were worn down and the sand from their erosion, which had been the major source of sediment for over 100 million years, gradually turned off. Then the major sediments came to be calcium carbonate, the residue of countless trillions of animal shells. Calcium carbonate (limestone) from Cambrian descendants is the major sediment in shallow, clear tropical seas today. If the seas were much more extensive then, as they seem to have been, the general climate must have been comparatively much warmer than today. The sea retains heat much more than land does and, with much of the earth's surface under water, a more tropical climate would be inevitable.

Seventy million years of life history following the Cambrian period lie grouped together under the name Ordovician, after another locality in Wales. The Ordovician beginning is marked by the sudden disappearance, for unknown reasons, of two-thirds of all trilobite species. Yet the group made a comeback and many new species appeared in the shallow seas widespread on all present continents. A primitive coral also began, as did the first jawless fish, the first snails, and the nautilus type of cephalopod. Cephalopods, represented today by the squid and octopus, were very common, suggesting that they had little competition. The shell of one was more than 10 feet long, by far the largest animal that had yet come along.

These new and old creatures lived in a world that was for the most part quiet. The North Atlantic coast of America, however, at times bristled with volcanoes during the Ordovician era. Volcanic ash from eruptions was carried as far west as Iowa. This seems odd because the winds must have been the very opposite of winds prevailing today. The Appalachian area where the volcanoes occurred must have been far south of their present position and in the tradewind latitudes where the winds blow from east to west. Data from ancient magnetism places the Ordovician Equator as moving away from its Cambrian position to just the degree where Appalachia would lie in the trade-wind zone.

At the same time that there was volcanic activity in eastern North America, there was vulcanism in the Caledonian mobile belt between Scotland and Norway. And at the same time both regions were undergoing mountain building. (In America this has the name of the Taconic Disturbance, after a mountain range in New York State.) The similarity of events in the two belts suggest that they actually were united in the Ordovician period, that the region only split later through continental drifting.

Another suggestion from the Ordovician that continents have moved about comes from the Sahara desert. Several years ago Rhodes Fairbridge of Columbia University, after an examination of rocks in northwest Africa with an international team of geologists, announced that the Sahara region had an Ice Age in the late Ordovician. The South

Pole then lay in what is now the hottest region in the world. This has been disputed by other geologists but it is an audacious and pleasantly unsettling idea. (Our old world seems capable of presenting us with all sorts of interesting surprises about its unstable past.)

Parenthetically, during the Paleozoic era a Siberian continental drift also happened and regions once together moved as much as 900 miles apart, creating an open sea where the Ural Mountains now exist as the Europe-Asia border.

With the coming of the Devonian "Age of Fishes" predators cleared the sea of the once dominant trilobites. The trilobites had survived for more than 300 million years but, poor creatures, now they were gone. The agents of this first major extinction on earth were the sharks and bony fish with good jaws and teeth.

With the advent of the Devonian period 400 million years ago, life, in the sea at least, began to be recognizable. Familiar forms appear; fish—you might say—about whom one might care.

The earth has been the theater of many great revolutions and nothing on its surface has been exempt from their effects. —John Playfair, Illustrations of the Huttonian Theory, 1802.

4

The Devonian period, usually given the name "The Age of Fishes" might equally well be called "The Age of the Backbone." For the later development of human beings, the evolution of the backbone, the quality of being a vertebrate, seems the most crucial Devonian event. With the support of a backbone, creatures late in the period even began to invade the land.

The earliest ancestors of the vertebrates are unknown and may always remain so if they had no hard parts to preserve. An axiom of evolutionary thought, however, is that no special creation exists; no species develops out of context. By logical necessity vertebrates had to develop from earlier forms, possibly from some distant relative in the echinoderm family, a family whose modern relative is the starfish. (No one proposes that we have evolved from starfish, however.)

The first vertebrates to leave a record were jawless creatures with the beginnings of a backbone, a brain, and rudimentary sense organs. Shaped somewhat like eels, with a cuplike sucking disc for a mouth, one descendant of this unattractive creature is the lamprey, the scourge of Great Lakes trout and whitefish today. Of course man is not descended from a lamprey, either, but may be more closely related to the second vertebrates to appear, These were armored, primitive fresh-water fish, not more than a foot long. The armor was protection from the water scorpions that lived in the same habitat. Some of these scorpions reached lengths of almost 12 feet. As is the evolutionary custom, the species of fish grew larger through the ages, faster, and began to lose their armor while, at the same time, the scorpions dwindled and disappeared. Many of the fishes of this fresh-water group migrated into the sea.

By the process of natural selection, fish later began to develop jawbones and so, in Devonian times, these "true fishes" became much in evidence. With jaws, fish did not need to live on tiny, bottom-dwelling organisms but could eat larger animals, including other fishes. One Devonian genus "Dinichthys" grew to a length of 30 feet. It looked like a large catfish with an armored head.

With the final development of an internal backbone by the "true fishes," an enormous advance had been made in the progression of living. Such an internal skeleton is more efficient as a strengthener than the external armor of the trilobites. It is also lighter. An external skeleton limits growth and, to overcome this problem, animals with such a burden must shed it every so often and grow a larger one. The constant manufacture of new skeleton uses up a good deal of animal energy. And while this new outside skeleton is growing, the animal is without defenses. With the skeleton buried inside, growth is not interfered with at any time.

Vertebrate development now branched out in several directions. One group of fishes began to lose the bony structure that had been developing and reverted back to a skeleton made of cartilage. These were the early saltwater sharks, one kind of whom, the genus *Cladoselache*, has been frequently found in the deposits that formed in the deep seas which covered the region that is now Cleveland, Ohio. These early sharks had teeth and, like modern sharks, could replace a tooth that was lost, perhaps by accident in the pursuit of prey, in a matter of 7 or 8 days. Sharks seem to have found the best solution of any animal to the dental problem. (Human beings certainly did not evolve a very good answer to the problem of wear and tear on teeth.)

The modern style of bony fish, the ones with which we are familiar, appeared and quickly rose to a position of importance in the Devonian lakes and streams. The pattern of bones in these ancient fish can be traced, although with modifications, in all high and later vertebrate types. Almost every part of the skull of a human being can be compared directly with a similar part of the skull of these ancient fish. Later animals have discarded some of the bones of the early fish but hardly any new bones have been added.

Apparently all the early bony fishes had lungs, although few modern fish still use them functionally for breathing. An explanation for the development of lungs may have to do with the most likely Devonian climate. This was a time when the seasons changed quite violently. Like some tropical conditions today, there would be very heavy rainy seasons followed by periods of great drought. If the ponds and streams where these fishes lived began to dry out, the water would become foul and lose the oxygen necessary for breathing in the water. If a fish had some sort of a lung sac, when the pond was drying up he could come to the surface and get the needed oxygen from the air, rather than the water as was his usual custom.

The first lungs were double sacs lying in the underneath chest region of the fish. This is the present arrangement in almost all land animals now but retained by only a few species of fish, ones that live in tropical regions subject to seasons of drought.

For most fish, when the climate became less variable after Devonian times, the lung served no function and even became a handicap, since a lung on the bottom of the body would make a fish top-heavy when it was floating. These fish modified the lung so that it lay above the throat rather than under it. Its use changed to that of a swim-bladder, similar to the ballast tanks on a submarine. The fish developed the habit of filling the sac with a gas they secreted, to the point where the animal's density was the same as the water density at the depth in which it was swimming. It could float without either rising or sinking. The sharks did not develop the lungs which later became swimbladders so sharks, for all eternity, must keep swimming or sink to the bottom of the sea.

Of the fish that did not give up functional lungs, three species survive today. Two of these, the lungfish of the Gran Chaco swamps in South America and the lungfish of the upper Nile Basin in Africa can exist even if their water dries up completely. They have learned to "estivate," hibernate in summer, by burying themselves in the mud, wrapping their tails around their faces, and sleeping until it rains once again. African lungfish have become so dependent on air breathing that if one is forced to stay under water too long it will drown.

The other fishes that kept lungs which functioned had a higher destiny than simply learning how to bury themselves in the mud. These other fishes are the ancestors of all land vertebrates today, naturally including the human race. These ancestral fish did not passively hide until the drought went away; they crawled over the land until they found water once again. (Several modern fish still move about on land; the climbing perch of India, the mud skipper and the walking catfish that recently became resident in Florida.) Many of the first fish who responded by crawling no doubt perished but enough were successful to keep this line of development alive. In this they were aided by lobed fins, with bones and muscles, that had first been found useful for propelling this fish on the lake or ocean floors; walking, in a way. These lobed fins could be adapted into limbs without any major change in the design. Nothing new had to be invented, merely altered. From these versatile fish came the amphibians; the oldest amphibian fossil that has turned up so far was discovered in Greenland. Its calculated age was Upper Devonian, about 350 million years ago.

The styles of fish that remained in the sea had to share their Devonian habitat with many competitors, including sharks and squid, both very aggressive animals but never a real threat to the dominant line of fish development. In later periods the true fish also had to contend with invaders from land who took up a marine form of existence; reptiles such as turtles and crocodiles, mammals like whales, seals, and porpoises, and the whole tribe of ocean-going birds. More often than not the fish were the prey of all these newcomers but nevertheless they thrived and varied to exploit new ecological opportunities until now there are about 25,000 species of fish. Most of them, both fresh and saltwater, are classified as the teleost fishes, from the Greek word *teleos*, meaning "complete, perfect" and the word *osteon*, meaning "bone."

The teleost fishes found success in part through a staggering variety of specializations. Although fish behavior is almost entirely instinctive, the signals for appropriate behavior are transmitted by a large battery of sophisticated sense organs. Fish may be stupid but they are very hard to fool. They have all the sense organs of mammals, plus a few extra ones.

Most fish have good eyesight and can see color but, as every diver knows, it is difficult to see very far in the water. Yet the elaborate color schemes of tropical fish suggest that they use color for finding the proper mate and perhaps in recognizing their own species in schools. Color is also used as a camouflage and sometimes to mimic a predatory species. Looking like a more dangerous animal would put potential attackers on their guard. Many fish also develop silvery scales; mackerels and herring, for instance. In the water these scales can act as reflecting mirrors so that the fish itself is invisible. Most fish are darker on the top and lighter underneath. Seen from below the fish is indistinct because it blends with the light coming from the surface. Seen from above, the dark surface blurs into the darker sea water. Many deep water animals, such as the lantern fish, have developed actual lights where it is almost entirely dark. Some lights attract prey, other lights are used to locate them, and the lights probably have a use in finding a mate. Mates are not abundant down in the abyss where food is hard to find. A number of deep-water fish are blind but some of these have lights and it is a mystery what use lights are to them. Presumably they are a vestige of the time when the species was not yet blind.

Such lights are an extreme adaptation of what is called bioluminescence, possessed by a great number of fish and other sea creatures. On land the ability to display lights belongs to glowworms, fireflies, and bacteria. The female glowworm is wingless but her light attracts the male, who only has a feeble light. Both female and male North American fireflies flash lights at intervals of about two seconds and this is a mating signal. In the Far East, male fireflies gather in great swarms and pulsate in unison, but the females take no part in such community displays and their meaning is unknown. Neither is it known why many kinds of bacteria emit lights. Much research has gone into the function of bioluminescence in the sea and it is now believed that it serves in mating, hunting, and controlling schools but there may be other undiscovered purposes.

One mystery about fish vision and color is the difference between the vivid coloration of tropical fish and the usual grayness of fish living in colder water. (It might be wondered why there are such distinct cold- and warm-water forms since, the world ocean being one body of water fish could swim anywhere they liked. The ocean temperature is not at all homogenous, however. It has very distinct temperature barriers and, once a fish is adjusted to one climate it will avoid any great variation.)

The brightness of tropical fish is most evident around coral reefs and here all their stripes and rainbow hues act as camouflage. In the tropics there are a great number of species but comparatively few members of any one of them, so the enormous variation in color would be useful in finding the appropriate mate.

In colder waters there are few different species but vast numbers of any one kind so that finding the proper mate would not be so difficult. Under the cold, gray skies of northern water a gray body would be an advantage. Parenthetically, tropical fish have a shorter life span and much less tolerance for variations in the water temperature. As an explanation for this, it is presumed that species differentiated in warm waters and that only the hardiest individuals could survive when, for one reason or another, they found themselves in colder water. Cold-water fish, as the rules of evolution say, are on the average larger than tropical ones. An extreme adaptation to the cold are the various ice fish of the Antarctic who, in adapting to a life of extremely low temperatures, have given up hemoglobin, red blood corpuscles. To live in this environment, these relatives of the trout have become virtually bloodless. To survive with a very limited supply of oxygen, these fish hardly move at all but the waters in which they live have a very abundant food supply.

Bloodlessness seems a very exotic development to mammals such as ourselves but surely of equal strangeness are all the fishes who live in an electric world. A vast number of electrical experiments have gone on in the waters of the world. The most spectacular result of these experiments must be the six-foot long electric eel of South American rivers. A true eel, it has the special ability of stunning a prey as far as 20 feet away by means of a strong electric shock.

Many different kinds of fish have, independently of each other, found ways of using electricity for avoiding obstacles

37

and for finding food. The electric impulses come from muscle tissue that has become specialized in various ways to produce electric impulses. Such fishes as can transmit electricity usually have special features in the skin, particularly around the head, which can receive impulses that bounce back off an object. The principle resembles that of radar. Electric fish have usually evolved in turbid water; in conditions of poor visibility, the ability to locate a prey in this way would be a great advantage. In addition such fish hunt at night, when most predators that hunt with their eyes are asleep on the bottom. During the day electric fish hide in inaccessible places, giving them considerable advantage in a highly competitive environment.

Fish do have ears but they are internal and not used for hearing. The ears' function is to balance the fish in its three-dimensional environment. Human ears do this, too, and using them for hearing developed after the ancestral species came onto the land. Fish "hear" by sensing through organs all over their body the vibrations produced when a sound is made. Salt water conducts sound five times more rapidly than sound moves in air, and all sorts of fish produce sounds themselves. Presumably the signals are communications concerning danger, mating, schooling, the discovery of prey, and so on. The noisemakers, catfish, groupers, snappers, and the like, are all carnivorous. Herbivorous fish do not respond to sound stimuli.

Numerous species of fish have developed chemical languages, based on extremely sensitive organs of smell. Some fish establish territories in which they are dominant and an intruder learns from chemical signals that he has trespassed. Hierarchies are maintained by the sense of smell. In an accidental experiment in a laboratory a large bullhead jumped out of his aquarium tank and into the one next to it which contained a number of small bullheads. The large bullhead attacked the little ones so severely that all but two of them jumped out on to the floor and died.

The two survivors, when rescued, established separate territories at opposite ends of the tank. To test their chemical memory, scientists put some water into their tank from the tank of the bullhead who had attacked them. The small fish panicked when they sensed the bullhead's secretions and, though the bullhead could not attack them again, the fish still fled and hid four months later when they were exposed to water that had been in touch with the attacker. Water from other bullhead tanks did not disturb them.

In further experiments concerning the role of smell in the social hierarchy, a dominant bullhead was put into the tank of a larger one certain to establish himself as superior. Returned to his first tank, the once-dominant, now defeated fish no longer caused the others to flee. His changed status was recognized by his altered chemistry.

The chemical language among fish is probably quite complicated. The odor from the fish's skin advertises his species, status, sex, age, and readiness to mate. Perhaps there is chemical communication between species.

Similar chemical sensitivity probably explains the homing instincts of the salmon, his amazing ability to return to the same little stream where he was born, after an absence of several years. Living their entire adult lives perhaps a thousand miles out at sea, the male and female return unerringly to their birthplace to spawn and, in the case of Pacific salmon, then to die. These handsome, powerful creatures have amazed men for hundreds of years as they watched them overcome even waterfalls to fulfill their instinct to reproduce. How do they find the right stream among all those along the Pacific coast? It has now been discovered that each stream has its own chemical individuality; no matter how slight the difference may be, the newborn salmon remembers it even after the long absence and always returns to it. Astonishing, but at least comprehensible as far as the river is concerned. Yet, can the salmon sense his own river even a thousand miles out at sea when its special character must have been diffused to almost nothing? Recent ideas on migrations in animals suggest that some do it by using the earth's magnetism as a guide. The magnetism varies from place to place on the earth like the differing features on a map and the animal could learn in his first migration the particular pattern he would need to follow to chart a return course. It is certain that some birds do this and since birds come from the same ancestors as fish, why should not fish also be equipped with magnetic perceptions as well? (It may even be that humans can sense magnetism to some remnant degree.)

Only Pacific salmon die immediately after spawning. Atlantic salmon do not, but return year after year to the same stream. Might the difference be that the Pacific watercourses are so much more difficult to negotiate that the salmon die as a result of their extreme exertion? Perhaps, but Pacific steelhead trout, which also return from the sea to their birthplace to breed, do not die as a result. Studies of Pacific salmon show that they begin to die as soon as they leave the ocean. The pituitary gland grows to about four times normal size, the metabolism speeds up, and the fat in the salmon's body is burned to almost nothing. The sleek, vigorous animal deteriorates completely within two weeks. How can evolutionary theory explain such a single-minded, destructive pattern for reproduction? Too specialized a behavior, one which allows no alternatives, often leads to extinction. The pattern for the magnificent Pacific salmon seems to be of this kind, accelerated by mankind's building dams as obstacles below almost every stream the breeding salmon must climb. Perhaps the salmon have turned into a dead end and the species is doomed.

Curiosity about the evolution of fish behavior is a fairly recent phenomenon and the field remains wide open for new discoveries. One large question concerns why some species travel in schools and how the individuals keep ranks so perfectly. Fish congregate in schools as soon as they are

40

born, still in the larval stage, and those which do not congregate cannot survive. Schooling may act as some kind of defense but, of course, a large school of fish is a more conspicuous target for predators. Schooling fish are mainly plant eaters so that this behavior cannot be a means of offense, such as that of an attacking army, but perhaps fish advancing along a broad front have a better chance of finding good pastures. The discovery of a choice feeding ground would be instantly communicated to the entire group. Fish in schools maneuver without any leaders. Those in the front suddenly find themselves in the rear, if the entire body wheels for some reason. The fish, almost all the same size, space themselves very carefully; they are repelled when they come too close to one another and attracted if at some distance apart. Control for these motions may lie in the "lateral lines," a system of nerves and sensory receptors along the backbone. Schooling probably also serves the mating function. When hormones announce that the time is ripe, schooling fish have no difficulty in finding an appropriate partner. One can only conjecture about the geologic age in which fish began schooling. Fossils do not tell much about social behavior.

Fish, some Japanese believe, may have the ability to predict earthquakes before they happen. In a book called "Fish and Earthquakes," Dr. Yasuo Suyehiro, director of a marine aquarium near Tokyo, wrote, "Fish can see ultraviolet rays that are invisible to the human eye and hear sounds that are inaudible to man. Although ultraviolet rays have nothing to do with earthquakes, it may be assumed that fish detect other signs of an approaching earthquake which humans cannot." The reason for such conjecture is that fish behave in very unusual, uncharacteristic ways before earthquakes. Before the big quake on May 16, 1970, an enormous cuttlefish that normally lives only in remote deeps was caught in a shallow bay just offshore from Japan. There are 127 well-verified examples of this kind of behavior. Presumably fish can sense profound activity beneath the ocean floor long before human beings realize any sign of warning at all.

Among the wonders of fish, their streamline design has been a major factor in successful living. Streamlining shows natural selection as a very superior engineer, the principles of which have been widely applied to automobiles, airplanes, and ships in search of speed. The barracuda, a creature not much praised by humans, nevertheless holds the speed record among streamlined fish. The barracuda can sprint at 22 miles an hour. Divers who have encountered them under water marvel at their ability to appear and then disappear too fast for the eye to follow. (Porpoises can sustain as much speed as a barracuda for much longer distances, but they, we remind ourselves, are much higher in the vertebrate scale, and are actually mammals themselves.) Evolution has given fish many others tools as well as speed for offense and defense. In addition to camouflage and electricity, many fish have extremely sharp teeth, others exude poisons; the scorpion fish excels here by having its poison at the end of protective barbs all over its body. In the struggle for life, evolution's unconscious ingenuity has created such strange situations as the remora, the pilot fish, that secures itself to the shark and is carried along to enjoy the fruits of the kill; the cleaning fish that set up stations where they safely eat other dangerous fishes' skin clean of bacteria and other growths, fish that would normally consider their cleaners legitimate prey; animals that have developed weapons such as the swordfish; animals that can blow themselves up to inedible size when threatened; animals like flying fish that simply leap out of the water when under attack. In the deepest waters of the world, the frightening-looking angler fish have widely extensible jaws so that they can swallow prey much larger than themselves. The reason for this must be that the population is so scarce at the bottom of the sea, dinners so hard to find, that the angler fish have adapted to make the most of what is available.

No species in life can prosper unless it mates successfully. More of nature's ingenuity has gone into the evolution of sure-fire breeding habits than into the solution of any other problem. Like the stickleback, in some other rare fish species the male cares for the young. The sea horse male (it *is* a fish in spite of its unusual appearance) seeks out a fertile female who will deposit her load of eggs in a pouch, like that of a kangaroo, on his stomach. The sea horse then seals the pouch with a sticky secretion and swims up and down in his usual manner for six weeks until the eggs are hatched. The grunions of the California coast wait for the monthly high tides to wash them ashore, where the female lays her eggs in a hole she has dug with her tail. There the male fertilizes them and the eggs are covered with sand. The ceremony always occurs at the highest point in the tide so that waves will not wash them away. The eggs hatch just at the next monthly high tide and the young are then carried back into the sea. One may wonder at such a procedure; how did grunions develop the instinct to obey the summons of the moon which influences the tide and learn just when to deposit their eggs on shore, safe from all the enemies in the sea. Californians make a sport of capturing grunions while they lie, helplessly, reproducing on the beach? Will the grunions with their inflexible habits succumb before the predators from land?

One of the rules about evolution observers have noticed is that a new species is most likely to evolve when part of a species becomes isolated. On its own, so to speak, an isolated group can develop along some particular line without its new specialty being overwhelmed by the common heredity, which tends to be conservative and rejects innovation. It is easy to see how a group could develop peculiarities if, say, a small number became stranded on an island. If a dominant male should have some peculiarity, a greater than average size perhaps or a slightly different color, he would very likely pass his difference on to many of the next generation. But how can a species become separated in the ocean, which is everywhere so much the same?

The apparent answer, learned in the past twenty years, is that the ocean is a much more varied habitat than it looks to be. It is much more nutritious at the mouths of rivers or where cold water, full of minerals, wells up along the west coast of continents. Cold water also contains more oxygen. The water becomes colder or warmer at various depths; the Atlantic having seven different layers of temperature. In some regions the water is fairly still; in others it sweeps along in strong currents. Thus fish are isolated in ways invisible to us and so evolve in different ways.

In connection with breeding a very common evolution among fish is to insure survival of the species by laying such an abundance of eggs that some must surely survive to become adults. The codfish, which spawn in the cold waters off New England, lay 6,500,000 eggs apiece every season. If out of such abundance only two eggs from every female survive to maturity, the species' security is assured.

Early curiosity about the breeding habits of fish led to some strange ideas about the behavior of eels. Eels can be found in rivers all over the world, but they are particularly common in Europe where they have been most often observed. European eels were never seen to breed and no one had ever seen eel eggs or larvae. The least wild suggestion about this was that they rose spontaneously out of the mud where they made their home.

When people began actually to look at nature, rather than just speculate about it, it was seen that some mature eels migrate down to the river mouths every fall and every spring young eels swim up the river into fresh water. Eels stay in the river for eight to twelve years, growing until the male may become a foot and a half long. When sexual development is complete for one generation of eels, they swim down to the sea and an estimate of 25 million has been given for the number of eels that depart from Europe every year. The natural presumption was that they migrated to some nearby coastal region for breeding. Yet no breeding eels or their eggs or larvae had ever been found in these waters, which were very thoroughly fished. No one had even seen the eels once they left the river mouth. They vanished. What became of all those eels?

A clue to the mystery came when an Italian biologist took a common little North Atlantic fish, of no commercial value, called a "thinhead" and put it in a laboratory aquarium. The scientist was amazed to see the translucent animal, which had much the shape of a leaf, slowly change shape as it grew and become very obviously a baby eel. (One of the difficulties about knowing how many species there are is that the young and mature of some forms are so different that they are classified as two separate kinds.)

"Thinheads" being at last correctly identified, the breeding place of adult eels still remained unknown. A dedicated Danish oceanographer named Johannes Schmidt made the problem part of his life work. After much time at sea and many years' labor, Schmidt plotted the occurrence of young eels of various sizes and finally drew a map which showed them all becoming smaller as the plot approached an area in the Sargasso Sea southeast of Bermuda. Taking a research ship there, Schmidt found a variety of eel eggs and knew he had discovered the true breeding grounds. But in addition to eggs from European eels, he found North American eel eggs! Both species used almost exactly the same area for reproduction. Why the same area, why did both species go so far, and how did they find it? (Presumably chemically, the way the salmon does.)

45

Later Schmidt looked for the breeding ground of Asian and East African river eels and found it south of Sumatra, above a deep ocean trench.

The chemical memory of fish explains how the eels find their way, and return to the correct continent, but why did they develop such a complicated procedure? The best answer today shows the strong relation between biological and geological events and demonstrates how animals may respond to changing environments.

The increasingly popular idea of continental drifting, the various demonstrations that Europe and North America were one land until 200 million years ago and then began to split apart, leads to a very likely explanation of the eels' breeding pattern. In the Triassic period, a new small ocean was forming where the original continent was breaking into two. Eels, which must have been saltwater fish since that is where they breed, invaded this newly forming ocean. Then some adventurous eels began to exploit the new environment of rivers, matured in the rivers, but retained the powerful instinct they had to breed in the sea. (Many kinds of fish have changed from one kind of water to the other.)

Their ocean birthplace, to which they had to return, became farther and farther away from the rivers as the Atlantic Ocean widened during millions of years. The separation of continents came slowly, perhaps two inches a year, and thus generations of eels would have time to adjust to the greater and greater distances they must travel before the female could lay her eggs and the male fertilize them. The mature European eels die after breeding, perhaps simply because of old age. Their young are born as orphans and grow slowly during the three years it takes them to reach Europe, unaided by parents. When they return to the Sargasso Sea as adults, twelve to fifteen years later, they must find the breeding ground using the chemical memory that was imprinted so long ago. But how as infants do they know where Europe is?

The thought that eels evolved their special habit because of a growing Atlantic Ocean is reinforced by the fact that their immemorial bedroom is not very far from the Mid-Atlantic Ridge, the line where the original continental separation began. The region in the Asiatic seas where Johannes Schmidt found the other great eel breeding grounds is over a 16,000 foot trench that is probably the crack where India, Southeast Asia, and Australia broke apart.

Geologic events such as the separation of continents obviously have enormous effect on the habitats of living things. And when land lies on the Equator it will support more different kinds of plants and animals than it will in polar regions. Biological happenings, the destruction of rocks by plants, the deposition of calcium carbonate that makes coral atolls, reefs, and even the whole state of Florida, these things clearly have a fundamental impact on geology.* Geology and biology are no more than parts of the same drama; with one subject, the natural world. Far too often the great story is broken down into fragments by specialists who forget the essential unity. The eels' behavior, always presented as a mystery, becomes no problem at all once it is related to the changing world in which they developed.

---

* Volume of the earth's continental crust has probably doubled since Pre-Cambrian times due to the addition of sediments that were organic (living) in origin.

From the summit of Mount Whitney only granite is seen. Innumerable peaks and spires but little lower than its own storm-beaten crags rise like forest trees, in full view, segregated by canons of tremendous depth and ruggedness. On Shasta nearly every feature in the vast view speaks of old volcanic fires . . . Southward innumerable smaller craters and cones are distributed along the axis of the range and on each flank. Of these, Lassen's Butte is the highest, being nearly 11,000 feet above sea level. Miles of its flanks are reeking and bubbling with hot springs, many of them so boisterous and sulphurous they seem ever ready to become spouting geysers like those of the Yellowstone.

—John Muir, <u>The Mountains of California.</u>

5

*S*ome natural events are swift. The destruction of Pompeii by Mount Vesuvius or the kill of an antelope by a lion are only moments in time. The epics, however, the creation of a dinosaur or the Andes, a South Pacific atoll or a golden eagle in flight, are spectacular changes that took ages to evolve. They are history in slow motion. One might wish scientists could say "On this day the world had its first giraffe" or "This was the year when North and South America joined together" but their tools are not sharp enough for such definition. It would be wonderful to see the great events in nature as they took place but the idea is only movie-scenario stuff. We can only look in amazement at the world we find and wonder how it came about. Surely the first explorers who suddenly came upon the Grand Canyon spectacle must have asked each other how such a thing as this could happen.

The Acadian mountain building in the Appalachians that ended the Devonian (and the domination of the world by the fishes) has been called rapid, abrupt, yet it took at least several million years. A recent theory is that the vast pressure which made mountains rise in Vermont came from a squeeze brought about by a collision between Europe and America. This joining together of two plates of the earth, which wiped out all traces of an older Atlantic sea floor was

49

just one episode in the giant motions that have taken place on our earth.

This collision is a place mark at the end of an epoch and the beginning of 70 million years during which the world's great coal beds were laid down. Europeans thus call it the Carboniferous but United States scientists break it up into the Mississippian and the Pennsylvanian, after the regions where rocks of the two periods are best exposed.

Until then the lands above water had been mostly barren rock with only small, primitive plants but, with geologic time-suddeness, the flora changed.

The plants had begun first in the sea, of course, and naturally they had to exist before the animals since the latter are, in a sense, parasites living off plant life. Flora in the sea discovered the magic of photosynthesis by which the energy of the sun is converted into growth. Animals never discovered this secret and must always remain dependent, receiving the power of the sun only at secondhand.

Most of the sea's plant life, phytoplankton, can only be seen through a microscope. The tiny, shrimplike organisms who are at the base of the animal food chain have the name copepods and they consume a large percentage of the plant plankton. Then, in turn, larger organisms eat copepods and, in the complex web of marine life, these organisms become the prey of even larger fish.

Plants more complicated than plankton can only live in shallow waters. At a time that will forever be unknown, some single plant cells settled on the rocks, and in a way not understood, began to attach themselves together, later growing into long threads, plumes, fronds, and all the other weird shapes of seaweed. These are not specialized cells in the sense of land plants. In the sea no need existed to form a tree with roots in the nutritious soil because the nutrients are around on all sides. The function of forming plant colonies, the reason for success, lies in the waves that pound on the shore. If a plant lies in still water it will deplete the supply of minerals and oxygen around it. If new water is constantly rushing around it, however, food and oxygen will always be coming along as fresh supplies. When the cells combine, their added strength allows them to keep in some position against the force of waves; unless, of course, the waves become too strong and tear them apart. The seaweed bends and gives with the waves. If it were rooted like a tree, waves would soon destroy it.

No one can do more than guess how plants made their way to shore. They had no hard parts to leave as fossils. Today no plant moves back and forth between water and land elements like the amphibians. One fancy has it that some ancient sea weeds did grow above the surface of the waves and thus became adapted to air, but no seaweed does this today. In any case we can see that they made it and adapted their inherited character to land conditions. Changes came about rapidly in the new environment. Plants developed outer coatings to prevent evaporation, learned to pipe up water from the ground and developed rigid structures so that they could grow tall, reaching toward the sun, and gaining height before competitors could put them in the shadows. Plant cells became specialized for absorbing water and minerals out of the ground, for capturing the light from the sun, for supporting the plant and for carrying water from below and food from above throughout the system. The engineering problems were many; undoubtedly for every successful experiment there were thousands of genetic changes that failed.

Plant and animal life of the Carboniferous period in Europe and North Africa was almost identical with that of North America, a fact that adds more support to the idea of close connections between the lands in the past. The dominant plants were large, scaly-barked trees and enormous ferns. The grounds were matted with creepers and vines. Later in the period, the gymnosperms, "naked seeds" appeared, known to us now in forms such as the

51

gingko tree and all the conifers—pines, cypress, and other evergreens. The creation of so many new kinds of forms in this new land way of life points up a very basic fact of evolution. Each new realm offers more chance for diversity and the total amount of living matter in the world goes on increasing, even though the environment already seems to be packed.

The climate of the coal-bearing regions, from all evidence, was very humid and very likely warm. Amphibians, who have no internal temperature control and thus cannot stand cold weather, thrived in the coal swamps. The fossil trees have no annual growth rings, which shows there were no important seasonal changes and the trunks lack the large, thin-walled cells trees have developed to counter cold weather. The vegetation of the coal beds certainly built up under standing water. The swamp water protected the vegetation from decay after it fell and then this matter was quickly buried under more vegetation in this rapidly growing forest. The present-day south Louisiana swamps, the Everglades, the Dismal Swamp and the Dutch lowlands have all been suggested as models of what the Carboniferous forests were like.

At the same time, while every northern continent had the same vegetation, all the continents in the southern hemisphere, as well as India and southern China, were dominated by a plant called *Glossopteris*. Leaves of this seed fern, which bore fruit, have been found impressed in the sedimentary rocks of southern lands now separated from each other by thousands of miles of ocean. The simplest way to explain *Glossopteris'* cosmopolitan distribution may be to say that the plant itself did not migrate and take hold all over the southern world, but that these lands were once all together in a vast territory now called Gondwanaland. (*Glossopteris* has been found as a major fossil in the Antarctic mountains and it seems very difficult to explain how this temperate plant grew there if the Antarctic has always had its polar position.)

The trees growing so abundantly in Carboniferous forests had large, luxurious foliage. In this primeval garden only amphibious animals at first stumped about. They still looked much like fish but they did breathe air. All early amphibians bear the name labyrinthodont, which refers to the labyrinthine folding of the tooth enamel. (Animal teeth vary a great deal between species and, since they are also the hard part most likely to be preserved, paleontologists know more about teeth than dentists do.) Labyrinthodonts, moving slowly about on lobed fins and, later, primitive limbs, seemed very clumsy, "looking like something a committee put together." Reconstructions show an animal that slightly resembles an alligator but smaller, stubby, without such a formidable mouth, less tail, eyes on the top of the head (as if they habitually lived three-quarters under water), and a third, pineal eye in the middle of the forehead. The short limbs were not directly under the body so labyrinthodonts would have had a very awkward walk. Most amphibians have only four toes, but some extinct species may have had as many as seven. (There seems to be no clear reason why higher species all settled for five toes as the proper number.) Their skulls were often armored, heavy and flat, and they presented a quite stupid appearance.

While labyrinthodonts moved about on land, they could never stray far from water, which they needed to keep their skin moist and which, very importantly, they needed as the place in which to lay their eggs. Turnover in labyrinthodont species was very rapid since they had entered revolutionary new realms in which to expand. Of course, their existence changed the ecology, too, and this required other adaptations. Evolution of some of these amphibious species finally produced a more efficient, four-legged land dweller but labyrinthodonts held on for many million years. Their oldest fossils appear in the northern hemisphere but, in 1968, part of the jawbone of a labyrinthodont was found imbedded high in the Antarctic mountains. It was at least 100 million years younger than the first fossil of his family. These ungainly animals had somehow managed to bridge the distance from Greenland to within a few hundred miles of the South Pole. The Antarctic labyrinthodont find has a greater significance, however. These amphibians were cold-blooded, could not have survived in such a rigorous climate, and would not have swum across more than 1000 miles of saltwater to reach the continent. A very good way to explain the Antarctic appearance must be to accept the fact that the animal was living in the region when it broke away from South America, South Africa, or Australia.

Edwin A. Colbert, the celebrated paleontologist who identified this amphibian from a small piece of jawbone, describes what may have been the end of these first land animals. He speaks of a find of labyrinthodont fossil bones in New Mexico. From the evidence he deduced that as a dry spell became more and more prolonged, these animals which so desperately needed water all congregated to what must have been one of the last pools in the region. There were hundreds, and from the bones, maybe thousands of them, in the pool "where this churning mass of amphibians came to its end."

What ambitious labyrinthodont, before they were all wiped out, made the momentous invention that has domi-nated all animal behavior ever since? When did this crea-ture begin to insert his sperm within the female to fertilize her eggs, rather than just spray his seed in the water where the eggs lay waiting? Like almost all the great revolutions in natural history, this discovery of a better way to insure species survival cannot be found in the fossil record. We will never know just how sex, in the way we understand it, first came about.

This new idea, the physical contact of male and female,

was strictly utilitarian. And it certainly began without plan. Some amphibious female was born with a mutation, a change in her genetic inheritance, by which she lost some of the terrible urge to deposit her eggs in the water, as tradition demanded. She must have simply dropped the eggs on land and a male, responding to age-old chemical signals, then tried to fertilize them. This, of course, would not work because the exposed eggs would dry out. How often did this happen before a male tried to fertilize the eggs while they were inside the female? When did the equipment necessary to do this evolve?

Idle conjectures, but certainly it did happen and with outstanding results. The fertilized egg now grew for a while before the female laid it and somehow a shell developed around the delicate material. This kind of egg had immense practicality. Rather than little jellylike globs of protoplasm which any lurking fish could devour, these first shelled eggs on dry land were safe from marauders since the amphibians still had the waterless region all to themselves. With a shell, the growing life had its own fluid and need not depend upon being in water anymore. It could survive droughts that dried up streams and could pass through the perilous tadpole stage of life in safety. Adult females could lay their eggs anywhere on earth that they liked. This opened up vast territories not necessarily convenient to rivers or lakes. With this new aptitude, some amphibians crossed a crucial biologic frontier. They became reptiles.

Of course the first species, called Seymouria, after a town in Texas where they were first found, still resembled amphibians in many ways. They did not linger long as a breed, disappearing in the face of competition from their descendants. In early Permian times, the north central part of Texas, the red beds where a great many fossils have been found, was a low, tropical delta. It received sands and muds from a large area to the north and east and was bordered on the south and west by a shallow sea. The delta had many streams, rivers, ponds and lakes. It was well covered with primitive plants and supported a varied, numerous population of both amphibians and reptiles strongly specialized for various ways of life. One of the largest Texas amphibians was Eryops, about five feet long, with a heavy armored skull, sprawling limbs, crudely articulated. The dominant reptile was Dimetrodon, a carnivore that grew to ten or twelve feet and had jaws with long, saberlike teeth. His victims were large fishes, amphibians, and even other reptiles.

Dimetrodon developed a large "sail" that ran down the middle of its back. It looked somewhat like an open fan and it puzzled generations of paleontologists. What function could this strange growth have. One suggestion was that the animal actually used the membrane as a sail and that breezes wafted the animal back and forth across Permian lakes. More recent opinion is that it was a method of regulating temperature, a large, four-foot high surface to absorb the sun's rays, and to radiate heat away from the body when it was too hot.

Insects had also made an appearance. Their ancestors were the arthropods, who also gave rise to the crabs and shrimps of today. There were cockroaches, spiders, scorpions, and one creature that looked like a centipede, except that it grew to be five feet long. The Permian fossil of a dragonfly has also been found. It had a wing-span of two and a half feet. Thus, the art of flying had already evolved once. How it came about no one will ever be certain but it is known that insects today are sometimes carried many miles by high winds and it may be that exposure to such conditions caused mutations toward movable wings to be advantageous. Men have no doubt very often asked why insects ever evolved at all? With a few exceptions, such as bees, they are of no use to anybody and creatures such as mosquitoes are plagues not only to human beings but to many other animals as well. The answer must be that the almost one million different species of insects that have existed evolved because there were niches available for them to fill. Specializing, finally, in being small they live in many spaces not available to any other creature. If some of them bite, this is both a good offense and defense. It does not seem to be the design of life that everything on earth survives only for the benefit of the human race.

During the hothouse period of the Carboniferous, the Equator ran approximately from what is now San Diego to Newfoundland. (In Europe, at the same time, it went from Denmark to central Turkey.) A month had shortened from 31.5 days in Cambrian times to 30 days in the Carboniferous due to the moon's motion away from earth. The land lay low and small enough in area so that rainfall was abundant in all seasons. Then this exceptional, lush period in the earth's history ended so completely that a different name for the following time exists in the geology books. This is the Permian, after a region in Russia, that lasted about 50 million years. During the Permian period the southern hemisphere had prolonged periods of glaciation. At one time the ice cap covered all of Africa that is now south of the Equator, including Madagascar which was still part of the continent. Australia, too, had an Ice Age with the ice coming north from Tasmania. Ice in South America reached within 10° of the modern equatorial line. All of India up to the present foothills of the Himalayas suffered from glaciation moving from the south. The distribution of these glaciers is a strong argument for a huge Gondwanaland that later broke up when the present southern continents began drifting north.

The end of the Permian marks the close of the whole Paleozoic era that lasted over 300 million years. This point in time was first noted by geologists because of the tremendous numbers of marine species that all at once became extinct. Among these were the last of the trilobites.

Climate changes of extreme severity were brought about partially by very extensive mountain building in the Appalachian area, in the far west of North America, and in the

Ural Mountains. Land animals and plants, as well as sea creatures, adjusted to warm, moist climates, faced "one of the great crises in the history of life." This Permian crisis caused drastic changes in all organisms. Either they adapted swiftly or they succumbed to the harsh new environment.

Hot molten material from the earth's interior rises to the surface in the middle of the ocean forming a feature known as the Mid-Atlantic Ridge. Once this molten material hardens, it is replaced by new lava from below, at the axis of the ridge, and the solidified rock moves slowly away from the axis in both directions—as if the blocks were on two conveyor belts moving in opposite directions.

Thus, the Atlantic appears to be growing at the expense of the Pacific because South America is overriding the Pacific. This hypothesis of sea-floor spreading was first suggested by the late Professor Harry Hess of Princeton in 1960. At first scientists, as they are naturally inclined to do, did not accept this concept. However, as often happens, measurements made in another connection shed new light on the hypothesis.

—From the Brunn Memorial Lecture given by Dr. Arthur E. Maxwell of the Woods Hole Oceanographic Institution at the Sixth Session of the Intergovernmental Oceanographic Commission, UNESCO, Paris, 1969.

6

*T*wo hundred million years, BP—Before the Present. The Paleozoic era had come to a disastrous end during the Permian extinctions a few million years before. In the sea, under the harsh change of conditions, most corals and echinoderms, almost all bryozoa, brachiopods, and cephalopods waned. Only one out of every four species of land animals survived. There has never been a more destructive time on earth. Yet, of course, many adaptable creatures did survive and went on to new specializations. The stage had been set for the Age of the Dinosaurs.

In this first stage of the Triassic period, the beginning of the Mesozoic era, all the continents were south and east of their present positions. To illustrate: by general agreement today, the 0 meridian of longitude is taken as an imaginary line that runs from North to South Pole, through Greenwich, England. If the same line is thought of as running through Greenwich in the lower Triassic, it would be 20° east of its location today.

The center of North America continued to be invaded by seas from the north and south. The waters waxed and waned as the world sea level changed and as the land rose and sank many different times. What is now the Pacific Coast was nothing more than an island arc of active volcanoes. To a degree, the position resembled that of Japan today, in relation to the Asian mainland.

Generally mild weather prevailed in the Triassic period and communication between widely separated regions must have been simple. Fossils of the same lizardlike creatures have been found in Africa, Brazil, Scotland, Western Europe, Russia, and India. This creature, with the difficult name of Rhynchocephalia, all but died out 50 million years ago yet a single species still exists today; the tuatara, the last of his race, can still be found hiding on some islands off New Zealand.

The Triassic period had an explosive development of various reptile types. Among the newcomers were turtles. They seemed to have appeared very suddenly but obviously they developed out of a long line from the Paleozoic. Turtles have been extremely successful animals, hardly changing their essential form for the past 200 million years. Numerous other dynasties lived alongside them, had dramatic rises and falls, but the turtles' way of a slow and steady life has been their sure road to prosperity.

Some Triassic reptiles took a very remarkable new course of action. It almost seems a reversal of evolution. A number of kinds of four-legged vertebrates, after millions of years of learning how to live on dry land, returned to a life in the sea. Perhaps driven by the competition, many reptiles became marine animals, some completely so, returning to the home of their distant ancestors. The icthyosaurs looked rather like dolphins and were streamlined for fast swimming. In time their original five-toed limbs became modified into fins for swimming. Ichthyosaur diet must have included belemnites, an early shellfish, because the rib-cage of one specimen contained 200 belemnite shells. Another reptile returning to the sea has been given the name of Plesiosaur. These strange sea serpents had bodies shaped like that of a turtle and a long, agile neck so that the head could move quickly to catch prey. Ultimately they reached a size of 40 to 50 feet. Icthyosaurs did not survive the Triassic period but Plesiosaurs lasted until the massive world crisis of 60 million years ago.

Flying was invented, for the second time, by some reptiles in the Triassic. The first innovators only managed to glide but ultimately the greatest of the pterodactyls, whose remains were found in Nebraska, achieved a wing spread of 24 feet, although its body was only the size of a wild goose. Pterodactyls probably soared over the oceans, like the albatross, in search of fish and seldom rested on land. Their final extinction perhaps came about because of their extreme specialization.

Leading up to dinosaurs themselves was the graceful, little lizardlike thecodont. His front legs were short and, when speed was needed, the limbs were lifted from the ground, and the thecodont ran upright on his strong back legs. He had a long tail to balance himself. His jaws had many sharp teeth so evidently his speed was used to catch small game.

Thecodont did not survive the Triassic period either but his descendants did so, exuberantly. The first animal called a dinosaur (from the Greek, "terrible reptile") was only a larger version of thecodont. What dinosaurs have in common, in relation to other reptiles, is that they were adapted to running. They carried their bodies off the ground, like mammals, and had their legs under the body rather than at the sides. There were two great orders of dinosaurs, having quite different origins and only remotely related.

Early dinosaurs were abundant in the eastern United States but their skeletons are rare because the dry terrain was not good for preserving bones. Anchisaurus, a prominent American form, was probably 5 to 8 feet long, and he left birdlike tracks, 3 to 4 inches long, all over the mud flats that covered parts of Connecticut at that time. These tracks became fossilized and only rediscovered in the last century. They show an amazing record of dinosaur activity, hurrying in search of food or resting between bouts of activity.

Dinosaurs, however, are not the only notable appearances in the Triassic period. The first hints of development toward the mammal condition of life began to appear. Some mammal-like reptiles, called Therapsids, have been found in a geologic formation in South Africa. These fairly large creatures had anatomies that suggest they might have had the very valuable ability of regulating their body temperature. They were the first true mammals, and are known as therians. They had rather small, rodentlike bodies; not very elegant to claim as our ancestors.

Out of nineteen major groups of reptiles and amphibians known to exist at the end of the Triassic period, nine suddenly disappear from the fossil record. The extinctions took place around the world. Why do such things happen? "After the Triassic the continents broke up and moved."

These are the words of S. K. Runcorn, a most eminent English geophysicist and one of the leaders of the continental-drift school of thought. Scientists often qualify such positive statements but not Dr. Runcorn. He has absolute confidence in his facts supporting this idea.

Such a breakup of Gondwanaland, and a Triassic-Jurassic date of 180 million years ago for its beginning, is supported by a mass of evidence found recently by drilling far into sediments on the deep sea floor. U.S. scientists have been studying fossil remnants of foraminifera and other microscopic creatures found just above the basement rock and finding that the very oldest forms cannot be called older than Jurassic, nothing yet older than 160 million years. We know that the seas are much older than this and that the remains of things living in the sea have much greater dates —but these dates are only for fossils now discovered on land. No one has been able to find any sea floor where the sediments above it go back any further than Jurassic. Yet we know the seas were there and that sediments must have rained on them throughout all the ages. Where did this older ocean go?

The new geology suggests that 200 million years ago all the world's land mass was together, nicely balanced north and south of the Equator. If there was some separation between northern and southern continents, in the region called the Tethys Sea, there was still enough communication between the two lands for animals to migrate back and forth.

Then some molten force within the earth exerted enough pressure on the surface to crack open this solid structure and send the various pieces on slow courses around the world. The thought is that the world broke into six major and a number of minor plates, crustal blocks of earth, and that these are the elements which drift. The force that moves the plates expresses itself in the rifts that make a more or less continuous ring two-and-one-half times around the world. At many places on this rift (which has a very high heat flow) lavas are coming out from the earth and displacing the crustal blocks to the right and left. These rifts are also centers of great earthquake activity.

The rift in the Atlantic Ocean is right in the middle, equally distant from the Americas on one side and from Europe and Africa on the other. The sediments in this center are extremely young but increase in age as the continents are approached. This was predicted before deep sea drilling began and has now been proven to be correct. The two oldest Atlantic sediments discovered were both near the U. S. coast. (The rift, of course, is the seam that broke open right at the beginning of this motion.)

In the Pacific the rift does not lie in the center of the ocean and this has posed problems to geologists. The rift, called the East Pacific Rise, lies near South America, and enters North America through the Gulf of California and, as it runs through the State of California, becomes known as the San Andreas Fault. Sediments on the East Pacific

Rise at sea are very young and increase in age as you drill either to the east or west. The oldest sediments found in the Pacific are near Japan but on the other side of the rift, the sediments do not become very old before you strike South America.

How does the theory explain this? It is that North and South America, two crustal plates that also include half of the Atlantic, have simply moved over the older eastern Pacific and destroyed all evidence of a more ancient sea. The ocean floor is not just simply crushed, as by a steam roller, but in a way consumed. This is easier to visualize in the case of South rather than North America, because the break in the North Atlantic came long before that in the South and so the collision between the South American and Pacific plates is much less advanced.

The west coast of South America has no continental shelf, no gradually deepening ocean floor leading out to the deep sea. The water just a few miles off shore is very deep and is called a trench. Only a few miles inland from the coast the Andes rise to majestic heights. What is happening is that the lighter continental rocks, made of granite, primarily, are riding over the heavier basalt that makes up the sea floor. This floor is gradually sinking down in the trench (at the slow rate of a few inches a year) and into the earth's mantle itself. Here it is melted by the intense heat created by radioactivity and often returned to earth again as volcanic lava. Much of the ocean floor does not melt, however, and simply by thrusting under them, makes the Andes grow higher. This is the cause of the earthquakes which are so common in Chile and Peru.

The same process occurred earlier in North America, first creating the Rockies and then later the Sierra Nevada. The continent's western edge now lies partly right over the rift, on the San Andreas Fault. The North American plate seems to be moving due west and the Pacific plate, which it is rubbing against, is in motion to the northwest. The friction between these vast masses creates the California earthquakes.

The old ocean floor on the Asiatic side of the Pacific does not collide with Asia because it is consumed in the trenches, the deepest cracks in the earth's surface, that ring the Pacific arc of islands such as Japan.

The same kind of collision that created the mountain ranges which run all the way from Alaska to the foot of South America have operated, since the continents began to drift, to create the Himalayas and all the mountains of southern Europe. India, after breaking loose from Africa, slowly moved north until it collided with the solid mass of Asia. No trench existed here and so all the tremendous amounts of material could only pile up until they created the highest mountains in the world—with sharks' teeth near the summit. The African plate, similarly shoving against Europe, has made the Alps and is presently elevating the mountains of Greece and the Near East. (Turkey and Iran have very frequent earthquakes.) It is presumed that Australia, New Zealand, and Antarctica reached their present positions through the operation of similar forces.

The effects of all this on biological life must at certain times have been catastrophic. Continents became islands. Lands once in equatorial regions suddenly, in geologic terms, lay in polar latitudes. New mountain ranges change the pattern of winds and sometimes create deserts. Death Valley, in the shadow of the Sierra Nevada, gets no rain because the high mountains just to the west condense all the moisture in the rain-filled clouds blowing from the Pacific.

A hiatus in the fossil record between the Triassic and Jurassic times had been known long before the new geologic theories. All the continents seemed to have sunk a bit and so the seas spread once more over low-lying lands. Restricted land areas reduced the ranges for land animals and presumably increased competition. This seems a ready explanation for the fact that some species failed, but why not all of them? From what can be told such a long time away, some species that disappeared seemed quite as likely to survive as those which actually did so. Nothing has baffled scientists more than the problem of extinctions. It is not that they cannot think of any reasons for a line to fail; there are far too many possibilities and many of them seem equally likely.

Yet, in any case, the Jurassic was a distinct era, quite completely dominated by all the dinosaurs. The early Jurassic was a very watery world for animals, with low continents and broad seas. Not in many millions of years had the seas covered so much land. What is now the western mountain region of North America lay beneath the sea and so did the Gulf Coast and the region now occupied by the West Indies. North and South America probably had no connection. Europe was nothing but a group of islands. Asia was an island continent and much of North Africa, New Zealand, and Australia were flooded over. For reasons that are not quite clear, fossil records early in this period are very thin. However, enough bones of carnivorous and herbivorous dinosaurs have been found in England, many of them imbedded in the chalk cliffs, to know that the trend was well underway toward giantism.

Later in the period the seas receded a bit, at least the Sundance Sea did so in the western United States, so that dinosaurs could range widely in the area and practically litter certain spots with their bones. The geological sequence in which these bones were found in the decades right after the Civil War was given the name Morrison Formation, after a town in Colorado located in the foothills of the Rocky Mountains just west of Denver. The Morrison Formation appears as bands of sediments along mountain uplifts in Wyoming, Idaho, Utah, Arizona, and New Mexico as well as Colorado.

Into this region about 1870 came two rich young American scientists whose feuds were almost as famous as their spectacular finds. On fossil hunts in the East the two had been friends but news of the Morrison Formation put an end to that. O. C. Marsh of Yale has been described as "coldly shrewd, crafty, grasping, wily, and pompous." His rival, Edward Drinker Cope was "ebullient, charming, brilliant, daring, and pugnacious." Both had great physical courage, tenacity, egotism, and talents for scheming. Since Marsh and Cope each wanted all the fossils in the West for himself, a fight was inevitable. They stole from each other, falsified, hijacked trains, robbed digging sites, bribed, slandered, libelled each other, went to court, and once their two digging crews engaged in a deadly pitched-battle over a site of dinosaur bones.

Marsh did make such famous finds as Brontosaurus, Diplodocus, the armored Stegosaurus, and the great horned dinosaur, Triceratops. These he gave to the Peabody Museum at Yale and the National Museum in Washington. Cope's major scientific achievements were with fossils of a much later period but his collection included many Triassic animals, too. Most of this went to the American Museum of Natural History in New York. In all, the Morrison Formation has yielded 69 different species of dinosaur.

The discovery of these unsuspected giants from the past caught the public's imagination. Newspapers carried sensational stories about the finds, and they were a source of a great deal of science fiction. Dinosaurs were so marvelously strange, so theatrical. The knowledge of these ancient creatures, which no human eye ever beheld, made many people realize for the first time how much drama there had been in the world long, long before man made his appearance. The existence of all these vanished species may have made a few people humble about their importance in the vast scheme of things. The dinosaur finds certainly made it easier for the idea of evolution to be popularly accepted, and they stimulated generations of scientific research.

During the time of Jurassic dinosaurs, the Morrison Formation was a vast alluvial plain, full of rivers and swamps. Its colorful purple, red, green, and gray shales today suggest a former climate that was tropical and humid. No fossil plants remain but vegetation must have been lush for the land supported herds of gigantic, plant-eating dinosaurs. Dinosaur bones are so abundant in the area now that a sheep herder was found by one of the early exploring parties to have built a cabin from them. These prehistoric fossils were the best available building materials. Some of the bones belong to animals that were sixty and seventy feet long and some of the creatures weighed more than fifty tons. Brachiosaurus may have reached eighty tons. The sauropods, vegetarian dinosaurs, all had heavy bodies, thick legs with broad feet, long necks and tail but with tiny skulls, not very strong jaws and pencil-shaped teeth. One ungainly sauropod, stegosaurus, had two upright plates along his back and spikes on the end of his tail.

The sauropods had carnivorous enemies, the carnosaurs. These were aggressive dinosaurs with enormous skulls, which gave room for long jaws and teeth like scimitars. (The Morrison Formation in addition contained crocodiles, turtles, lizards, frogs, and some very small, quick mammals.)

Dinosaurs in the Triassic period had become vastly specialized since their first reptile ancestor laid an egg. The triceratops found by O. C. Marsh is a good example of how far they had come and suggests some dinosaur dilemmas created by all this apparent progress.

Triceratops grew to a length of 25 feet. He stood heavily on his four feet, had a short, stout tail and a huge head up to 8 feet long. The head supported a bony frill that led back over the neck, two large horns that pointed forward over the eyes and a shorter horn on the end of his nose. He needed very strong neck muscles to support this great skull. The horns were presented to the attacker in defense. Triceratops probably also used his armament in combat with others of his species during fights in the mating season. Presumably the females, being reptiles, laid eggs in holes they had scooped out in the warm sand.

The animal was cold-blooded and not being able, like mammals, to regulate his internal body temperature, he could only exist where the temperature did not vary much. With his enormous bulk he surely had to eat vast quantities of food. From the evidence of his teeth, paleontologists say his diet consisted of coarse plants that grew on land. His brain was so small that he must have been extremely stupid, yet the species, protected by their size from all but the greatest carnivores, survived well enough in their ecological niche as long as conditions there did not radically change.

Children are ever delighted when they come upon the huge dinosaur skeletons in museums and certainly the sight of them is a very graphic way to get a sense of the long time of earth history. But of what use are dinosaurs to us? The last of them disappeared millions of years before the first creature resembling a human walked the land. They offer no threat or competition to any living thing. Why should anyone care about dinosaurs; go to the enormous labor and expense of excavating them, studying them and reconstructing their skeletons? Just for amusement?

The complete extinction of the dinosaurs has worried generations of scientists and no one yet has a satisfactory answer to the question of how such a dominant group could simply cease to exist. Speculations about the extinctions are very numerous but the simple fact is that no one knows. Yet to understand extinctions is important, if the history of life is capable of telling us something important about the future of our own species. As Edwin H. Colbert, famous paleontologist, writes in *The Age of Reptiles*,* ". . . extinc-

---

* Edwin H. Colbert, *The Age of Reptiles*. W. W. Norton and Co., New York. 1966.

tions are evolutionary processes, just as are origins. There would be no evolution without the origin of species, which replaces the old with the new, and at the same time there would be no evolution without the extinction of species, which removes the old to make room for the new. The world is now a very different world from what it was some seventy million years or so ago, when the last dinosaurs were at the pinnacle of their power, and *it is different to a very considerable degree because these dinosaurs became extinct.*"

Yet the Jurassic graveyard of the Morrison Formation was occupied long before the end of the dinosaurs' reign. They still had their zenith to reach in the forthcoming period.

And Jurassic rocks had another record of a great event in life history. While it seemed apparent that birds must have evolved out of a line from reptiles still the first bird fossils discovered during the great Victorian hunt were quite far advanced, a considerable remove from their presumed forebears. A missing link was required. Then in 1860 a quarry used for lithographic limestone in Germany yielded the impression of what would have been called a small dinosaur except that it distinctly had feathers. Teeth also existed in this mold but they were considered accidental because everyone is aware that birds have no teeth. The discoverer named his fossil archaeopteryx, or "ancient bird." This find was sold to the British Museum and then a better-preserved archeopteryx was found in Germany a few miles away from the first discovery. The aggressive O. C. Marsh from America tried to buy this but German nationalism was aroused and it went to the University of Berlin for $4750. A third, inferior specimen was found in the same region in 1956.

Feathers function not only as a means of flying but for keeping the bird warm as well. (Birds evolved their efficient thermostats independently of mammals). A frequent problem raised in evolution is that something like a feather is far too complex to have arisen in one or two mutations, but at any intermediate stage it could not have performed any function. How does a perfect feather suddenly appear? The answer is that it does not. Feathers, which are a development out of scales, were first a selective advantage because they conserved heat and only later did they prove useful in flight. Whether archaeopteryx had warm blood is not known but it is known that later birds did, and that the development had nothing to do with the same change in mammals because the blood circulation is quite different. Neither one of the systems could have evolved from the other.

Discovery of Archeopteryx, the perfect connecting link between reptiles and birds, intensified the question of how the first birds started to fly. How did they live to make this an advantage. Two theories have been proposed. One was that certain lizard types took to the trees to nest, feed, and avoid predators. A wing membrane could have developed that would help the animal jump from branch to branch, as the flying squirrel glides today, and that eventually this could have been adapted to powered flight.

The other idea was that a wing membrane helped in swift running. In time this grew stronger and larger until eventually the animal found it could take off and glide for a few feet. At last full-fledged wings grew out of this new ability.

Recently a scientist from Yale found a fourth archaeopteryx, quite by accident. While looking around at the fossil collection of a museum in Holland, he came upon a specimen supposed to be a small dinosaur but it did not look like that to him. Tilting it up to the light he saw small feather impressions in the limestone. It had been found in Germany before the discovery of Archeopteryx in 1860 but it was a rather poor specimen and no one knew then what to look for.

This fourth bird, somewhat more primitive than the others, had a horny sheath around the claws found on the bird's wing. It did not seem useful for clinging in trees but would have been of advantage to a predator on the ground. Thus, at the moment, it looks as if birds learned to fly by first running along the ground.

After the time when these first known birds flew in Europe, the record of their development hardly exists for millions of years. Their bones were too light, their bodies too fragile, for good preservation and by their habits would seldom be caught in the muds or asphalt beds that have saved other animals to become fossils for the paleontologists. Birds did not become prominent until the new period began, the Cretaceous, "the time of flowering."

The train south to Lake Titicaca and Bolivia leaves Cuzco at 7:30 in the morning, climbing gradually onto the puna. This land much resembles part of Wyoming and Montana and, higher still, the foothills of Mount McKinley—the same lupine and bunch grass, dwarf willow and alder, of our alpine tundra. But there are many unfamiliar flowers, including a lovely species of silver blue, in clumps, and jucca cactus, and the sagelike tola, and representatives of bird families not usually found at these altitudes in North America—coot, and a variety of hawks, and white egrets, and the serranita gull, and, most astonishing of all, their color reflected in a long, shallow lake beneath snow peaks, a flock of forty or more flamingoes. The pink of these strange birds as they rise against the distant snowfields is an ornothological incongruity beyond description.
                                    —Peter Matthiessen, The Cloud Forest.

7

When the world had reached an age of 4,400,000,000 years, the flowers finally came. Magnolias may have been the first of all. Magnolia leaves appear in a Greenland deposit which dates from the beginning of the Cretaceous period. This period name comes from the Greek "creta" which means chalk and also applies to the chalky island of Crete. The British conceived a new use for the word when they discovered that their white, chalk cliffs are actually made from the calcareous remains of once living plants. Coccoliths, microscopic in size. Yet if you did not know what Cretaceous really means, you might easily be persuaded that it meant something flowering, blooming, bursting with life. Purists object to the word "explosion" for an event that made no sound, but the period was the time when flowering plants rapidly took over the world.

Before flowers hit the land, its cover was a monotonous green. Previous to the Cretaceous trees were mostly cycads, conifers, and gingkoes. Cycads are extinct now in the northern hemisphere with the exception of Central America and Mexico. The primitive gingkoes, once common around the world, had retreated to Japan in modern times but were brought to Europe in the eighteenth century and then to North America where, particularly in New York City, they seem to thrive in spite of the climate and the air. The conifers have held on better in the face of competition yet they have retreated from their once world-wide dominance as the newer broad-leaf trees have replaced them. A few tropical conifers exist today but mostly they survive in less hospitable, colder regions. The redwoods of California are part of the cypress branch of the conifer family and they are so specialized in habitat that their continued existence is threatened. (Even more so when a Governor of California can say, "If you've seen one redwood, you've seen them all." This for trees that have lived 4000 years.)

Sex in a new style was the revolutionary weapon of the new kind of plants. Botanists call the breed "covered seeds" or angiosperms; they had a different design for fertilization. Before the Cretaceous period, most plants, called gymnosperms or "naked seeds," simply reproduced by loosing a mass of spores onto the world (the way most fish do) in hopes they would be fertilized. The new seeds came enclosed in an ovary with a sticky-ended projection called a stigma. When a pollen grain touches the stigma, it stays fast and quickly grows a shoot down into the ovary to fertilize the seed. Plants and animals, independent of each other, both evolved the same sophisticated method of sexual reproduction.

The plants employing seeds gained a supremacy over those using spores because seeds are stored with food reserves for the growth of the young plant within. Once fertilized, little is left to chance. The seed is so cleverly constructed that once it is lodged it needs no water but can immediately root and send out leaves. Of all seeds, nuts are the height of achievement, so far, in this kind of development.

The well-stocked seeds of angiosperms do not travel but wait within the parent for the much smaller, active flagellate, the sperm, to approach and fertilize. Yet the sperm cannot itself move to the egg and must rely on some other power to transport it. And here is the strange accommodation. A plant flowers not to aid in growth but to spend itself in providing for the next generation. Its pollen grains wait for a messenger to take them to another flower of the same species. Somehow, first birds and then, more importantly, insects became attracted to flowers. The plants' offer in exchange for this service is a color, scent, or nectar to please the visitor, who will inadvertently carry the pollen to the waiting egg. How such a marvelous thing came about in the first place can never be known. Since the first events, of course, both plants and insects have specialized in thousands of different ways so that certain plants, for example, only attract certain insects. Waste is reduced and the chances of a correct mating much improved. With the advent of flowers of many kinds, the numbers and species of insects also increased tremendously. The flowering plants, (almost all the plants we know, enjoy, and depend on for food) were a development in evolution whose significance can hardly be overstated.

Flowers came during a period when the seas had advanced over the continents to a degree that had hardly ever been exceeded in the past. There seems to be no particular dramatic reason for the large marine invasions all around the world. Apparently the lands had been eroded almost to sea level and thus a slight rise in sea level could have a very broad effect. Shallow water covered the bulge of Africa. Britain, Denmark, and much of northern Europe received sediments during much of the Cretaceous period and the seas spread into western Siberia. North America was about forty percent submerged and the continent was actually two large islands. Debris from the western island poured in great quantity onto the low-level area—a geosyncline—that existed where the Rocky Mountains now lie. This region kept subsiding as more and more material was added. For some of the time, however, parts of the region supported coal forests and Colorado today has extensive coal fields laid down in the Cretaceous period.

During the period, the lands gradually began to assume the positions they occupy on modern maps. Africa drifted north about 10 degrees of latitude. India was moving north across the Equator. Australia and Antarctica still touched each other. Eurasia was rotating clockwise and this motion, along with that of Africa, began to close the eastern end of the Tethys Sea. Greenland still touched Europe but the Atlantic Ocean was steadily widening.

During this span of about 60 million years the dinosaurs thrived and evolved into many bizarre new species. The first true snakes appeared. Marsupials did, too; animals like the kangaroo and opossum who carry their young in pouches. An event of considerable importance to life as we know it was the arrival of the first placental mammals. This is, of course, the line which led to the primates and the human species but we would be hard put to it to guess this outcome from the first little insect-eating placentals.

The world climate of the Cretaceous period was generally very mild. Footprints of a dinosaur, which could not endure cold weather, have been found on the island of Spitzbergen, 10 degrees north of the Arctic Circle, where the warmest temperatures today are barely above freezing.

Then two major events marked the end of the Cretaceous period and the beginning of the Cenozoic era. These were the Laramide Revolution and the Great Death. The Laramide Revolution, named for mountains in Wyoming, marked the birth of the Rocky Mountains and the death refers to that of the dinosaurs. It is not true that "the Rocky Mountains killed the dinosaurs" but the two matters coincide in time to some extent.

Many parts of the world began to feel crustal unrest as the Cretaceous period ended but none more so than the western part of North America. Where the seas had lain so long depositing their sediments now became the stage for mountain building, folding, and thrusting, on a colossal scale. The movement ran all the way from Alaska to Central

America and in the United States region the band of disturbance was more than 500 miles wide. The Laramide Revolution was one of the great facts of geologic history.

The mountain-building forces came from the west and such was their power that Colorado had at least one mountain, Long's Peak, thrust up nearly as high as Mount Everest, 29,000 feet. (Erosion has worn most of this down.) There was much volcanic activity during this thrusting and every western state, in what is now the United States, had active volcanoes.

Creation of the Rockies was no sudden affair. The process took some millions of years. How could it have affected the dinosaurs? In any event, though the Rockies have been blamed, the dragons became extinct not only in North America but all around the world. Fifty different reptile families lived during the late Cretaceous period and only fifteen lived on into the Paleocene era 75 million years ago. Five of these vanished in the following Eocene epoch and only snakes and lizards were left. The Great Death not only occurred on land but in the sea and included not just seagoing dinosaurs but a dominant shellfish called ammonites.

No single theory seems to explain the great extinction. All the dinosaurs, large and small, herbivorous or carnivorous, died out. Yet lizards, snakes, turtles, and the early mammals did not. Flying reptiles vanished but birds thrived. Dinosaurs vanished from the sea but fish and sea turtles carried on. The extinctions cannot all have come about because of competition. Nothing large enough to overthrow the dinosaurs existed and it took the mammals millions of years to occupy their vacant places. Something world-wide is needed to show why all sorts of genera, many families, and whole orders all died out at once. It can hardly have been pure coincidence.

A plague has been considered but rejected. Certainly no one disease could wipe out so many types of creature completely and leave so many others intact. A suggestion has

81

been made that races can die out because of "old age" but no one has been able to show such communal exhaustion and, in any case, why would it overtake many different races at the same time? Over-specialization has also been blamed but is it plausible in so many different cases at once? It has also been said that dinosaurs became too large, but the many small dinosaurs also vanished. (And they had developed into giant sizes and thrived that way for millions of years.) One theory has it that climatic changes brought about the fall of the dinosaurs but, in fact, Paleocene climate was little different than the Cretaceous, and why should climatic changes not affect every kind of species? The rise of flowering plants has also been given as a cause but they had been in existence for a long time before the mysterious disappearance and there is no reason to suppose dinosaurs did not find the plants an excellent source of food. It has been supposed that perhaps the new mammals stole and ate dinosaur eggs and probably this is true, but eggs are always being stolen, very commonly today, without wiping out so many kinds of life simultaneously.

All of man's probing still has not turned up any better evidence to support an explanation of the extinction of the great dinosaurs. Edwin H. Colbert, in his *Age of Reptiles,* noted: "It was an event that has defied all attempts at a satisfactory explanation, for which reason, among others, it has fascinated paleontologists for decades."

Chances are best that it was a combination of factors that finished off the Age of Dinosaurs. We should probably be very happy that they are gone because, without them, the mammals had a chance to thrive and take over dominance of the world.

8

*The last era of the time lead-*ing to Homo sapiens has the name of Cenozoic, or Tertiary, a name covering about 75 million years. Epochs within it, each a bit more than 10 million years long, have fluid names that for easy recognition are much too like one another. Some rather glib diagrams try to make it all simpler by assigning a characteristic to each of the six times. Thus, Paleocene has diversified hoofed mammals. Eocene saw modern types of mammals. Oligocene the first running mammals. Miocene, the first abundant grazing mammals. Pliocene, large carnivores. Pleistocene, early man and continental ice sheets. Such simplification makes it sound as if evolution had devoted a certain number of years to one production, then turned attention to some other development, until finally the world all became neatly arranged for the triumphal coming of man. But of course all life was evolving at once. Each form we know today can be seen in the beginning of the Paleocene era and its development followed in considerable detail. It is not too elementary, however, to call the whole era the Age of Mammals. This great group now includes almost all the largest animals on earth, as well as the whales, seals and porpoises, and bats, who fly.

Although we all know perfectly well what a mammal is, a

look at the definition could be salutary. After a particularly depressing news broadcast, it might help to remember all the wonders that have been wrought since our species developed from a single-celled species in the sea.

Mammals nurse their young. Most birds care for their fledglings, too, but few lower vertebrates do this. Some primitive mammals still lay eggs but none with which we are familiar. Usually the embryo is fed within the mother's body for a long period of time. Mammals regulate their temperatures with hair, sweat glands, and other, internal mechanisms. The circulating blood in healthy mammals is kept at a very exact, optimum temperature in spite of externals.

Compared to other animals, the brain of even the most stupid mammal is enormous. The teeth are highly differentiated; at first most suitable for a carnivorous life but varying later in adaptation to various eating habits. A. S. Romer suggests that mammals can be distinguished from more primitive animals by a one-word summation, "activity." The first mammals were meat-eaters and they needed speed to capture their prey.

Species most likely to be successful are, naturally, those whoch do the best job of insuring the survival of their young. All the listed mammalian developments contribute toward that end. In addition, the direction has been toward fewer offspring, but larger ones. Long after their first appearance, mammals remained small and primitive. Perhaps any larger ones who might challenge the dinosaurs were quickly eliminated. When the dominant reptiles vanished, for whatever reason, the mammals had had a long apprenticeship, specializing in better brains, and they were ready for the new opportunities available.

Early in mammal history there may well have been as many pouch-bearing marsupials as there were placental mammals. The pouch into which the very small, helpless kangaroo crawls after birth has obviously been a fairly good

84

method of nourishing the young, but the placenta seems to be a better one. The protected embryo has a long period of development before birth, more protection, and placental infants receive much more training after birth. A placental infant can function more efficiently as an adult than the marsupial infant which had to develop a degree of independence, at the expense of reaching complexity, at an earlier stage.

In nature only two ways to make a living have ever developed among animals. Either the animal lives off plants or the animal makes his meals off plant-eaters. (Obviously, the most successful animal, man, owes much of his success to the fact that he did not specialize in this way but will eat any living thing in that world that is not poisonous—including artichokes.)

Humans may well wonder why any species would become a plant-eater and subject to the sudden death shown so graphically in movies of lions hunting in Africa. Why would any animal choose to be a victim when it could have become King of the Jungle? (Of course this is anthropomorphism, against which people are so sternly warned, because it prevents understanding of how things actually work in the animal world.)

Species can only develop out of their past and what they may become is determined by what has happened before. An insect with an external skeleton cannot readily change its predetermined size. A bird cannot offer mother's milk to its young. The emerging species of mammals all had many differences; not vast differences but enough to strongly influence the direction of their evolution.

The first mammals probably were insect-eaters, carnivores of a sort, but they very likely could digest some kind of plant food. If insects were in short supply, some would doubtless begin to specialize in plants.

Those who became primarily carnivorous are identified today mainly by their teeth. Teeth were and are their principal method of killing and these teeth need to cut through tough hide, tendon, and bone to get to the flesh. But flesh is not hard to digest and does not need much if any chewing. Thus, animals that are primarily carnivorous have lost their grinding molars. Cats cannot chew at all. Members of the dog family, not so exclusively meat-eaters, do have all but one pair of their grinding molars and can do some chewing with their cheek teeth. Bears have turned away in general from the meat-eating habit and their teeth have been modified so they can chew well. Proper carnivores also developed supple bodies and claws to attack and hold their prey.

By Oligocene times flesh eaters had evolved into two main groups that resemble the modern carnivore families. For simplicity, one could say that one family includes the cats while the other family contains the dogs. Surprisingly the large cat group includes the Old World hyenas and such small, tropical flesh-eaters as civets and mongoose which never got to the New World. (Included here are the exotic carnivores of Madagascar, all primitive civets who became isolated when the large island separated from Africa long ago.)

True cats are called felids. They are not long-distance runners but stalk and then hit their prey with a sudden jump. Felids include the domestic cats, lions, and tigers. Though lions and tigers look quite different with all their fur, it is evidently impossible to tell them apart by a comparison of their skulls. Another large cat as big as a tiger, the saber-tooth called Smilodon, roamed in North America until the end of the Cenozoic era and then became extinct, perhaps due to extinction of the mastodons on which it preyed.

The group of carnivores to which dogs belong includes many species particularly familiar to North Americans. Among these are weasels, martens, fishers, skunks, wolverines, badgers, and the otters who took up fish-eating and their relatives who went to sea.

Wolves, foxes, hyenas and even coyotes are closely related to dogs. Konrad Lorenz, the celebrated Austrian who has written so well about animal behavior, considers from his understanding of the way they act that some domestic dogs descended from wolves and others come from a hyena stock. Certainly, domestic dogs and wild wolves have been known to mate successfully.

Raccoons evolved in remote times from the same family as dogs. The pandas and Giant Pandas of the Old World belong in the same line of descent. Bears are still another branch of this inventive line of carnivores but only one of them, the polar bear, is a pure carnivore today, perhaps because it lives in a land with nothing much to eat except fish. To make the list complete, carnivorous mammals also went back to the sea and became seals and the walrus. (How they solved the problems of mammalian reproduction in this new mode of life is a fascinating, but separate, story.)

To support a food chain there must obviously be very many more plant-eaters than carnivores. A lion, for instance, needs to kill every few days and will go on doing this all his adult life. The large quantities of prey that all these flourishing species certainly thrived on must, in their turn, have had very great quantities of food. The major source of this food in the Cenozoic era were the grasses that began to appear on the earth. Grasses—the forage plants—developed from the plants of the Cretaceous flower revolution. Later, grasses included cereals and grains, such as oats, rice, corn, and wheat, with all their consequences for civilization, but Miocene herbivores had already begun to prosper on the grasses that began to spread on the plains.

Scientists describe all herbivores as ungulates. This means that they are hoofed, although the class also includes sea cows and certain animals with claws. The taxonomy of ungulates (their classification based on their evolutionary history) seems rather artificial. Hoofed animals are grouped, for instance, into what seem unimportant categories; they are odd-toed or they are even-toed. Perhaps the history of life may make sense even if one does not understand why animals are divided in such a way.

The horse, most famous of the odd-toed ungulates, first appears as a fossil in Eocene deposits (50 million years ago) of the western United States. Presumably he began in North America and later made his way in successive invasions across the Bering Straits to Asia and then moved on to Europe. The original horse, now known as eohippus or "dawn horse," was no more than a foot high, about the size of a fox terrier, with a slender face and long tail. His legs were long and a reduction in the number of toes, from the ordinary five, had already begun. Eohippus had four toes in front and three behind. His low-crowned teeth suggest that he browsed on leaves in the forest.

Fossil horses of the Oligocene epoch are now called mesohippus. This ancestor had only three toes on each foot and he had become larger; perhaps the size of a collie. One suggestion to explain increasing size is that predators would be more likely to attack a smaller prey first and so there would be a selective advantage in bigness. Another thought is that large browsing animals can reach more food than small ones, as well as run faster. In addition, a larger animal can go without food for longer periods.

By the Miocene epoch parahippus and merychippus had begun to develop high-crowned teeth, presumably because they had changed habitat and diet. Now the early horses had moved out onto the plains and were eating the grasses that had just begun to change the barren plains into prairies. These horses still had three toes but only one touched the ground. This continuous development of the feet in the direction of a single toe made the horse an ever more

88

efficient runner, able to live in the open country and gallop away from predators, rather than hiding.

The changes in horse anatomy, as seen in abundant fossils, do not occur gradually but come about rapidly. The speed of the transitions probably means that intermediate types were inefficient and so selection quickly eliminated such specimens, leaving only the successful for the paleontologists to find. By the time of the Pliocene, a horse called hipparion had only a single toe and had reached the size of a pony. The modern horse, equus, found its way to Eurasia and flourished in North America as well, until suddenly and mysteriously it disappeared from the American prairies *after* the last Ice Age glaciers had retreated. No horses existed in the land where they had evolved, then, until the Spaniards re-introduced them to the South American continent about 1500.

Like many extinctions, the disappearance of the horse from the Americas has defied explanation. However, Charles Darwin himself did have some interesting comments on the subject in *The Origin of Species.* "The extinction of species has been involved in the most gratuitous mystery. Some authors have even supposed that, as the individual has a definite length of life, so have species a definite duration. No one can have marveled more than I have done at the extinction of species. When I found in La Plata the tooth of a horse embedded with the remains of Mastodon, Megatherium Toxodon, and other extinct monsters . . . I was filled with astonishment; for, seeing that the horse, since its introduction by the Spaniards into South America, has run wild over the whole country and has increased in numbers at an unparalleled rate, I asked myself what could so recently have exterminated the former horse under conditions of life apparently so favorable." Darwin went on to discuss the fact that there are unfavorable conditions which we cannot perceive, even though they exist around us.

". . . every creature is constantly being checked by unperceived hostile agencies; and these same unperceived agencies are amply sufficient to cause rarity, and finally extinction."

Darwin was not surprised that the huge mastodons and dinosaurs died out. ". . . as if mere bodily strength gave victory in the battle of life. Mere size, on the contrary, would in some cases determine quicker extermination from the greater amount of requisite food."

The prophet of evolution did not know exactly what happened specifically to the horse in the Americas but he had a general reminder for his readers. "The theory of natural selection is grounded on the belief that each new variety and ultimately each new species, is produced and maintained by having some advantage over those with which it comes into competition; and the consequent extinction of the less-favored forms almost inevitably follows."

Another odd-toed hoofed mammal originating in North America was, surprisingly, the rhinoceros. Light, running rhinoceroses and amphibious ones appeared in Eocene times. These had no horns and grew to about the size of their modern relatives, the tapirs. True rhinoceroses with horns came somewhat later and were very common on the Great Plains till they died out during the Pliocene epoch, when the climate was cooling. One rhinoceros that lived in central Asia about 30 million years ago, the baluchitherium, may have been the greatest land mammal that ever lived. It was 13 feet tall and had a length of 25 feet. The last remnants of this once-important group now live only in the tropics of India and Africa where they face the danger of extinction.

Finally, among the odd-toed ungulates, a vanishing race, one may mention the zebras.* Zebras once ran in enormous herds across the grassy plains of Africa, protected by their numbers and by their remarkable stripes that are presumed

---

* Even the horse, among the one-toed species, exists only at the pleasure of man. As a wild animal, the horse is practically extinct.

to be camouflage. A favorite prey of lions, they also became the target of men with guns and are now scare outside of parks and sanctuaries.

Among the surprising facts in the history of nature is that camels originated in North America. Camels also persisted in what is now southwestern United States until quite recent times. Of the group artiodactyla, even-toed mammals, the "cloven-hoofed" creatures, they began as small as a jack rabbit and such a fossil camel has been found alongside the skull of an eohippus. These first camels had no humps, a development that came only when they began to dwell in the desert rather than the plains. The hump, which permits the animal to go for days without water, and the broad cushion-like pads on its feet which help it walk without sinking in soft sand, are adaptations that give the camel a means of existing in regions without much or any competition. Camels are ruminants, members of the group with complicated stomachs whereby partly digested food is returned to the mouth, the "cud" is chewed, and then sent to another stomach to continue breaking down the roughage.

When South and North American finally joined together perhaps 5 million years ago, camels migrated south past the new Isthmus of Panama, settled among the mountains and became llamas, vicunas and similar forms of animal. Other camels crossed the Bering Straits and ultimately found their way to the Sahara Desert. The final skull of the original American species was found, with dried flesh still preserved, in a cave in the state of Utah where lava had flowed after the last Ice Age glacier. In modern times camels from Africa were re-introduced into the southwestern States and thrived in the deserts. Why they could not survive there after the Ice Age remains still another mystery.

Some very successful ruminants such as the deer and its related forms—elk, moose, reindeer, caribou, and such—have developed stomachs more complicated than that of the camel. Almost all of them have a taste for temperate or cold climates and they all have antlers. Antlers, while certainly weapons, should not be confused with horns. Horns remain permanently attached to their owners but antlers are shed every year, only to grow back in greater size and complexity. In terms of efficiency this evolution seems terribly wasteful since so much of the animal's food intake every year must go into the construction of new antlers. All of the deer family, so familiar in North America, are actually immigrants who must have come by way of Alaska. The only purely American ruminant is the prongbuck antelope which rose quite independently but nevertheless developed the same system of digestion.

The giraffe, a ruminant always certain to draw a crowd at the zoo, has made an interesting solution to the problem of keeping alive. The giraffe browses on leaves, rather than grazing on the ground, but he lives in savannas where trees are not so plentiful as in a forest. Therefore the long legs and long neck which make it possible to reach high branches not available to competitive animals. In spite of the long neck, the giraffe has only seven vertebrae to support it. Almost all mammals have seven neck vertebrae and not even the whale has departed from this number which goes back to the earliest forms. Fossil giraffes demonstrate that his long neck is a very recent innovation.

The greatest number of higher ruminants belong to the bovid family. Although a number of species now exist in North America, only three arrived here by migration across the Bering Straits, The buffalo, mountain sheep, and mountain goat came late in the Cenozoic period, when the climate was getting progressively colder, and evidently these were the only bovids who could stand the temperature. The group had their start very probably on the tropical plains of Asia and most of those that later adapted to frigid regions did so only because of domestication by man.

The most prominent members of the bovid family are cattle, yet in spite of familiarity today their past heritage is not well known. This applies equally to sheep, goats, and antelopes, who are also bovids. Zoologists classify them together because they all possess true, steadfast horns. They might equally well be grouped together because they have been conspicuously successful in the evolutionary struggle. Mankind, of course, has contributed to their prosperity because their stomachs are capable of digesting grass which men are not themselves capable of using directly.

When the classifiers of animals came to elephants, they gave up on odd or even toes. Elephants, although ungulates too, obviously belong in a class by themselves. Probably no one has ever called an elephant beautful but no observer would deny their unique appearance. Rudyard Kipling had one explanation for elephants but others have said that they also look like something put together by a committee. Trunks seem so bizarre to humans that they ask why in the world would any species invent an outrageous thing like a trunk. The obvious answer, clear to anyone who has ever seen an elephant pick up a peanut, is that trunks are useful. And so their scientific name is Proboscidia, after Greek words having to do with feeding. Once acquired, a trunk has utility but how did it come about in the course of evolution?

Ancestral proboscidians made their first home near ancient Lake Moeris, now in the country of Egypt, about 40 million years ago in the Eocene epoch. At present this is the Fayum Desert but then it was part of the Nile Delta, humid and forested. Here the early progenitor of the elephant, as yet without a trunk, also founded the lines that led to dugongs, mantees, and the rodent-like conies of Africa and Syria. Called Moeritherium, his incisor teeth are the clear beginnings of the later proboscidian tusks. Nature is full of strangenesses but there are not many facts stranger than that elephants should have come from the same stock that produced sea cows such as dugongs and manatees, the suckling water animals that produced the legends about mermaids. And how did manatees get to Florida and Brazil while dugongs found their way to the East Indies? These species must be less than 40 million years old, yet continental drifting began millions of years before that. The sea cows had to swim a long way from their beginnings in Egypt.

The relationship of these animals to elephants stems in part from the fact that moeritherium itself was amphibious. So we can understand the evolution of sea cows, and elephants are known to be extremely fond of water. A further oddity about this stock is that the word "coney" for the animal that developed from moeritherium is really an old English term for "rabbit." English translators of the Bible had no personal experience of this animal mentioned in the sacred book so they simply wrote it down as coney. And then, an oddity, we have Coney Island which really means Rabbit Island.

In the process of evolving to a larger size, the mastodons who followed moeritherium began to have difficulty reaching the ground for food. They could not develop longer necks because their heads became so massive that the necessary muscles for such a neck would have been mechanically impossible. In the species called Trilophodon the lower jaw grew progressively longer and the upper lip followed suit. The lower paw was used for rooting, scooping, or shoveling while the upper lip, more flexible, tucked away food and squirted water down the animal's throat. Eventually the upper lip progressed, became stronger, and more versatile, and the mastodon could use it for food gathering on its own. With this improvement, the lower jaw began to get in the way and began to recede.

Greater tusks provided an evolutionary advantage not only for offense and defense but for stripping the bark from trees. The tusks later seemed to have gone too far. In a creature named anancus they were so long and heavy that the animal must have had a difficult time moving its head at all. With dinotherium, the tusks curved backward. Some

94

later mammoths had tusks whose tips crossed, so they were useless except perhaps except as a guard for the trunk.

A theory about evolution is that a trend sometimes gets carried away with itself and specializes to a point where it becomes a burden that can lead to extinction. Thus, with anancus and the mammoths whose tusks crossed, it would seem matters had reached this point. There was no adaptive purpose, no function, in such developments. Yet each of these species survived for more than a million years. They persisted in spite of obvious handicaps.

How can this be explained? One answer lies in the idea that evolution functions to preserve the species, not the individual. Extravagant growths do not reach their climax until the animal is approaching old age. Tusks that might be extremely useful when the elephant was of breeding age might only become useless or dangerous after he had reached an age where his survival as an individual was no longer of any importance to the species. Anancus as a vigorous young male had long, powerful tusks of great use to him. When he grew older and his tusks became fatally overdeveloped, he was merely paying for the earlier advantage he possessed. Older elephants today still grow unwieldy tusks.

Fairfield Osborn of the American Museum of Natural History says there have been 352 different species of elephants. During their long history they grew progressively larger because it gave them a selective advantage. As with many other animals, size produces more strength and perhaps more speed. In colder climates, increased size produces a more efficient heat-regulating system.

Yet size can be carried to extremes. The mechanical cost of becoming a mammoth demands sacrifices in other developments. Being committed to bulk means that enormous quantities of food must be consumed to fuel the machine. Elephants must eat continuously for about 16 hours a day. Should the food supply diminish for any reason, they are in immediate danger. So much specialization makes them very much at the mercy of their environment. While the vegetation remains normal, elephants can thrive but lacking versatility they cannot survive much of a change.

To support his great frame, the elephant developed legs thick as pillars. This meant that he could not run very fast. Yet the size also meant that no adult elephant need fear an attack from any predator. No animal could grow large enough to present a challenge because such a size would require the intake of so much food that they would rapidly exhaust the supply of available prey.

So the elephant tribe was very successful until recent times. They radiated out from Africa in all directions and by Miocene times had crossed the Siberia-Alaska land bridge into North America. The woolly mammoth roamed North America in great numbers during the Ice Ages and many frozen carcasses of this animal have been found in the tundra of Siberia. At about the same time, the Imperial mammoth lived in Texas. Perhaps the tallest of elephants, 14 feet high, he had tusks that were 13 feet long.

Overspecialized as they were, the elephant tribe did remarkably well until a predator appeared who was undaunted by their size. This was man, naturally, who found the huge beast a target hard to miss and who, in any case, could outrun an elephant on an open plain. Elephants were hunted by the early Indians in North America and the breed finally became extinct here 8000 years ago. The once-lordly animal is now reduced to a habitat in zoos and circuses and roams free only in India and Africa. The Indian elephant can be domesticated but the larger, African breed for some hereditary reason will not tolerate a subservient position.

An ancestral gull population, possibly breeding in the Bering Straits, increased in numbers and colonized new areas both to the east and to the west. Although gulls have great potentials of dispersal, in practice individuals usually return year after year to breed in the same area, often in the same colony, as that in which they were reared. Consequently a fair degree of reproductive isolation would exist between the original population and the new colonies derived from it. This partial isolation would make possible some divergence in habits and structure. Finally, populations spreading eastwards must have met similar populations which had spread to the west, the meeting-place being on the other side of the globe in the North Atlantic. When this meeting took place, the two forms were sufficiently different to remain distinct.

—John Maynard Smith, <u>The Theory of Evolution.</u>

9

*G*eolologists *consider their subject* to be the earth and its rocks. Biologists deal with living things. Actually geologists and biologists are only looking at two aspects of the same dynamism and nowhere is this more obvious than in the case of South America. In South America geology, or geography, absolutely controlled the evolution of life for at least the 130 million years when the continent was an island. During all this long time the forms of animals developing, often by competition, in North America, Eurasia, and Africa could not reach South America and so the animals there evolved in comparatively peaceful isolation.

While the above ideas are not questioned by scientists, nevertheless there is disagreement between most of the younger ones and some of the older ones about the reasons for South America becoming an island. The differences depend on one's attitude concerning continental drift. As Bruce Heezen of Columbia University said, until 1955 any academic who spoke in favor of continental drift was likely to be condemned and ostracized. The reasons for there being such an emotional attitude about a purely scientific problem are not the subject here but such emotional attitudes do still exist. Having said as much, the assumption here is that drift has essentially been proven and it certainly

99

helps explain the peculiar prehistoric life of South America.

One historic reason for the idea that continents now far apart once were joined together is the very neat, jig-saw fit that can be made by fitting the tip of Brazil into the Gulf of Guinea in western Africa. Comparison of rock formations in Brazil and the African country of Gabon show they are almost identical. Fossils of the extinct plant Glossopteris exist on both continents now separated by several thousand miles of ocean. Permian fossils of the reptile Mesosaurus, the first land reptile to return to the water, are found on both continents and many of the ancient crustaceans were also the same.

Deep-sea sediments drilled by the U.S. ship *Glomar Challenger* show that the South Atlantic is between 130 and 150 million years old, so that is the general time when South America and Africa began to break apart. At this period the world-continent had primitive mammals not yet very well differentiated into the marsupials and placentals. Primitive ungulates also existed but had not begun to evolve into even-toed and odd-toed forms. Yet South America was already adrift and alone.

Most writers say that North and South America had been joined together before the geologically recent reconciliation. No particular evidence for this seems to exist but obviously South America once had connection with the rest of the world and, if continents were stationary, with whom could South America have been joined? If one accepts the union with Africa, however, North America is not needed and, in any case, the westward drift of North America seems to be some million years previous to that in the South Atlantic.

During South America's long time as an island many special animals appeared including carnivorous marsupials (not much of a success on the world-continent), toothless mammals such as anteaters, tree sloths, and armadillos, and various hoofed plant-eaters not immediately related to animals on any other continent. Some time later specialized rodents and New World monkeys became added to the animal landscape. Anti-drifters would have these animals arrive by some kind of "island hopping" from North America but this seems to be purely guessing and no one knows even if the necessary islands existed by which monkeys and rodents could or would make their way. (And no fossil monkeys have ever been found in North America.) It seems just as easy to think that these animals simply evolved independently but along similar lines as different species in the rest of the world. New World monkeys are certainly not on as high an order as those of the Old World and no good reason exists for their not having come about on their own.

Then earth forces in the Pacific Ocean caused a major revolution. For ages the waters of the Atlantic and Pacific had mingled freely. During the long Cenozoic era the chemical composition of the seas in this generally American region changed several times so that the bottom sediments of once-living microscopic animals changed from having shells made of calcium carbonate, to much harder shells made of silica, and then back again. Presumably a change in water temperature caused this but no one is certain. In any case, the ocean floor in both Atlantic and Pacific shows the same changes. About 10 million years ago an ocean ridge existed, from the present position of the Galapagos Islands, eastward along the Equator toward South America. Then because of some thermal action deep below the surface the ridge began to split into rift zones. The southern portion remained in place but the northern half split into several pieces that began to drift slowly northward.

A deep sea trench lay off what is now the coast of Panama and Costa Rica, a deep channel between Caribbean and Pacific. (The north part of the trench still lies off the Central American coast.) Piece after piece of the Galapagos ridge drifted into the trench and was pulled down until it finally engulfed and became closed. The reaction on the eastern side of the trench was a rise above sea level in compensation. In this way the Isthmus of Panama began and then later the southeastern part of Costa Rica. Some of Cost Rica is still rising today in the slow motion of this massive event.

Precise dates are not possible for such happenings and no one knows how long it took for land animals to discover the new territory and invade it. In any case by the early Pleistocene epoch, somewhere around one million years ago, animals from both North and South American had found the land bridge and crossed it, with tremendous consequences.

Motion came in both directions. Fifteen families of North American mammals spread into South America and seven families traveled in the opposite direction. From North America, rabbits, squirrels, field mice, dogs, bears, raccoons, weasels, mastodons, horses, tapirs, peccaries, camels, deer, and cats migrated south. For South American mammals the results were disastrous. Old hoofed animals were replaced by trespassers from the north. Many native rodents became extinct when rabbits, squirrels, and field mice took over their ecological niches. Placental carnivores wiped out the marsupial beasts of prey. The main survivors were armadillos, anteaters, monkeys, and such, who had no ecological competitors from the north. In all cases of competition it was the northern animal that survived.

Such a cataclysm certainly brings to mind "survival of the fittest." The North American animals had been exposed all through the Age of Mammals to the constant evolutionary changes taking place on the world-continent. Those who made it to the Pleistocene epoch were products of natural selection, found at the end of tremendous competition. They had learned to invade and to meet competitors. While competition had existed at first in South America, the victors had long since found niches that did not challenge other species and the economy was, simply, live and let live. When the threat came about through the closing of the Isthmus, these animals knew no adequate way of defense.

(What took place among the animals in South America can be easily compared with a great many episodes in human history. Consider all the nations that have been invaded and conquered. Painful and costly though a military defense certainly is, the pain and cost stemming from no defense may be even greater. Of course, argument about defense by means of analogy with South America can be made to seem ridiculous. South American had been isolated for more than 100 million years. How in evolution could animals have been prepared?)

Distance seems the obvious factor in whether or not an island will develop its own unique animal life. Great Britain is too near Europe, Ceylon to India, Japan to Korea and Siberia, for instance, and thus their animals are like those of the mainland. The strange island of Madagascar seems just crucially far enough away from South Africa and separated by a deep channel so that its animals are very different. Madagascar has no hoofed animals at all except a bush pig that probably arrived with immigrants from Malaysia. It has just a few odd carnivores. Many primitive primates such as lemurs but no monkeys, who appear later in evolutionary time. It had a number of ground birds, some very large; birds that gave up the flying habit because there were no serious competitors on the ground. The West Indies had no native carnivores but the islands nearest South America had snakes. Even though Cuba is only 90 miles from Florida it had no North American mammals except the aquatic manatee. Cuba is separated, of course, by the fast-moving Gulf Stream.

The islands of the Pacific have been particularly fertile grounds for ideas on the variety of life. Charles Darwin's thoughts on evolution came in great part from observing the various finches of the Galapagos Islands. ". . . seeing this gradation and diversity of structure in one small, intimately related group of birds, one might really fancy that from an original paucity of birds on this archipelago, one species had been taken and modified for different ends."

Fourteen species of finches live on Galapagos, about 600 miles out to sea from Ecuador. They vary little in their behavior during courtship, in the eggs and nests, or their anatomy, except for one major difference. Each species has a beak of different size and shape, depending on its feeding habits; whether seed, cactus, insect and so on. Sometime, perhaps driven by a storm, a flock of ordinary seed-eating finches arrived from the mainland. Few, if any, other species of bird lived there; hardly any competition. Responding to their opportunities the finches underwent "adaptive radiation," taking over roles that would be occupied by woodpeckers, warblers, and tits, if they had existed on the islands. In time the birds became so varied in their habits, depending on the chief source of food on each island, that they are now essentially different species. They are sexually selective and choose their mates according to whether or not they have the correct shape of beak.

The effect of isolation on animal life can also be seen in the marine iguanas on Galapagos. These iguanas go to sea in order to feed and they are the only iguanas in the world that do so. They have close relatives on the South American mainland but these relatives are not marine. Somehow, in the past, iguanas got to the Galapagos Islands. The land there contains very little vegetation so the lizards found their way into the sea and took up a diet of algae. When foraging, the iguanas can stay under water for an hour. While diving the circulation of blood to the muscles is cut off and blood only moves between the heart and the brain. (This same system is also used by diving mammalian seals.) The oxygen used by the muscles while swimming is drawn off the tissue and, when back on land, the animal has to rest for a long while, breathing and paying back its oxygen debt. These unique iguanas can be partly tamed and be taught to enjoy an omnivorous diet.

The islands of New Zealand were rifted off, from Australia, over 60 million years ago during the Cretaceous period. If any mammals existed before isolation, they were few and died out, and the only indigenous mammals there now are two species of bats which undoubtedly arrived by flying. Birds perhaps existed on New Zealand before separation and others arrived by the air route. Lack of predatory mammals permitted some birds to live a life of foraging on the ground, eventually becoming flightless.

Separation permitted two quite exotic types of birds to develop on these islands. Once there were 20 species of moas, flightless, and growing to more than 10 feet high, Maori hunters, immigrants of the Polynesian race, wiped out the last moa hundreds of years ago. The little nocturnal Kiwis, now symbols of New Zealand, thrived a bit better but are still difficult to find. Their distinctions: they lay an impossibly large egg and they are the only living birds who have nostrils at the tip of their long beaks.

Another relic of a very ancient past is the tuatara, a reptile of up to 2 feet in length that is not a lizard but the last survivor, the living fossil, of animals related to the dinosaurs. On every place else on the earth they became extinct millions of years ago. The tuatara is rare and as well protected, perhaps, as is possible for any government in a world where there are heedless people congenitally unable to leave harmless animals alone. Over countless ages, the tuatara has developed the curious habit of sharing nests with the petrels. The bird makes the burrow and then the tuatara moves in, apparently without giving the bird any concern. At the end of summer the petrel migrates and the reptile remains in possession. Tuataras live to extremely great ages, several

105

hundred years, and so they outlive the petrels and become sole owners of nests built by other hands. How such an arrangement came about will surely remain a mystery.

New Zealand possesses a number of other animals less exotic than these and in the last 100 years settlers have introduced many species, evolved in a more rigorous world, who seriously threaten the native islanders. Some American, for instance, apparently introduced speckled trout into a New Zealand mountain stream and this predator has flourished at the expense of the native fishes.

Perhaps the best studied case of island evolution may be Australia, because of its bizarre types. Drift theory has this continent attached to India and Africa until Gondwanaland began to come apart. Indeed, marine invertebrates found in Australia and dated Permian resemble those of the same age discovered in India and Africa. The separation of the Antarctic from Australia came at a later date. Australia went through the same kind of geologic history as the other southern lands, including various ice ages and invasions by the sea. During the Cretaceous period, Australia may have found something rather like its present position. For a time, then, it must have had a connection with Indonesia. Marsupial animals certainly made their way from there to Australia. For reasons that one can only speculate about, placental mammals did not make the trip or, if they did, they did not survive long. The two types of mammals came onto the world scene at almost the same time so it does not seem likely that prior evolution made Australia available to marsupials and not to placentals. Then the short-lived bridge to Asia disappeared beneath the waves and marsupials spread out in this land without competition, and began "climbing, jumping, running, burrowing." They became "herbivorous, insectivorous or carnivorous."

Sharing all this with the marsupials were quite a variety of exotic insects and two curious mammals having fur and nursing their young, but still laying eggs in the fashion of reptiles. The duckbill platypus has some teeth when young but none as an adult. It is a good swimmer and digger, living by the side of streams. The spiny anteater has powerful digging feet and a long slim snout to carry out its ant-eating activities. These weird creatures, called monotremes, probably preceded the marsupials but they were archaic, no challenge, and would certainly have never survived until modern times unless they had been protected by isolation.

Many of the Australian marsupials began to disappear immediately after man began to introduce the higher placentals and, though kangaroos seem to be holding on, it would seem to be only a matter of time before those animals which succeeded in the struggle on the world-wide stage will replace the indigenous animals, except as curiosities.

Among regions in the Pacific far too distant to have had any natural invasion of mammals,* the Hawaiian Islands are of interest touristically, agriculturally, and to oceanographers, volcanologists, and ecologists. In terms of ecology they are but another sad example of what may happen to a beautiful region when man comes with his weapons and his animals. A great number of bird species, for instance, have become extinct. Species who have been protected too long by the quiet ways of islands have a particularly tough time when confronted by the main stream of world evolution.

---

* Except one species of bat and one species of seal.

Please do not feed the Animals.

In 1965 the Council of the Society decided that the feeding of the animals by the public both at London Zoo and Whipsnade Park must be stopped in the interests of the health and general welfare of the animals. It was agreed that this policy would be introduced as new animal houses were opened, and at the present time the only important groups of animals which may still be fed by the public are the monkeys and apes, some of the bears, elephants at Regent's Park when in their outside paddock, and the herd animals at Whipsnade.

It has now been decided to carry this policy through to the final stage and feeding by the public of any animals or birds will no longer be allowed.

Much has been learned in recent years about the diets needed by individual species to preserve their health and enable them to breed under zoo conditions. One of the most important facts is that diets should be carefully controlled. Obviously unbalanced diets are inevitable if visitors are allowed to feed the animals, and this can lead to problems concerning nutrition and growth and can also cause disease.

Some animals are unable to discriminate in the choice of food offered to them and may become ill or even killed by eating quite small quantities of the wrong food. Others, diverted by the lure of food offered by visitors, develop the quite untypical habit of begging. This can seriously reduce exercise and activity, leading in some cases to minor digestive disturbances. More tragic consequences sometimes follow such as the death of the elephant "Diksie," which fell into the moat surrounding her enclosure while reaching out for a tidbit.

Last, but not least, the feeding of animals can cause injury to visitors and their property as the animal takes food from them. The elephants alone over the last year have seized no fewer than 14 coats, 12 handbags, 10 cameras, 8 gloves, and six return tickets to Leicester, damaging them beyond repair. In the same period there have been 180 incidents involving bites by monkeys, parrots, cockatoos, ponies, and other animals.

The Society has for many years protected its special feeders such as the Giant Panda, Giraffes, and Polar Bears, but modern zoo management is based on the necessity to control the feeding of all animals. Proper management means the creating of conditions in which animals may breed, since animals of many species are now available to zoo collections only from the surpluses bred in other zoos. The control of feeding in zoos is necessary to create conditions in which these animals may breed and rare species may be conserved.

—From a note to members of the London Zoological Society.

10

*A*t *the beginning of the Age* of Mammals, palm trees grew in Alaska and England and crocodiles thrived in New Jersey. As the era ended much of the world lay under huge ice caps. The Alps, Andes, Rockies, and Himalayas had risen to their awesome heights. Man, the hunter, was at large. By the time the ice finally retreated 11,000 years ago, man had become ascendant. He had the measure of every living thing. (With the exception of snakes, insects, and diseases.)

Dates for the first Ice Age and the first appearance of true man are not exact. Both dates keep receding on the geological timetable. All reckonings of time's extent have had to overcome the notion that the world was created in 4004 B.C. It seems now that true men and first great ice sheet came on the scene at about the same time. And the rapid evolution of man must be due in part to the challenge of the threatening ice. A clever, highly adaptable creature would surely take advantage of conditions that would only bewilder more stupid animals. In any case men certainly took over the caves which provided some comfort from the cold. They painted fine pictures on the walls and improved life by manufacturing stone tools.

The trouble with reconstructing a history of man detailed enough to show how he may have evolved in response to the challenge of the ice, in terms of body changes, is that so few fossils of early man actually have ever been found. It may be that he was actually too clever to get caught in swamps, drown in rivers, be trapped in quicksand or asphalt pits. Evidently early primates were rather smart, too, for none of them have left many remains. It may also be that primate remains are few because almost all were tree dwellers, and bones on the forest floor are quickly consumed by animal predators and by consuming, destroying protozoa such as bacteria.

Other than being generally dwellers in trees primates are difficult to define. This stems from the fact that they *lack* specialization. They retained the old-fashioned five fingers and toes, collar bones (which support sideways movement of the arms), and their teeth did not become so specialized that they had to settle for either being predators or plant-eaters. Primates, mostly living in the forests, developed good eyesight for judging distance while swinging around on trees. Their hands improved for the functions of grasping and holding on. In time flat nails took the place of claws. This permitted the tips of the digits to become very sensitive, better suited for manipulating objects, a considerable improvement over having to do such things with a mouth. In the forest, smell did not matter much, so the sense diminished in power and the nose grew smaller.

The oldest primate fossil so far discovered turned up in the same Egyptian fossil beds that yielded the ancestor of the elephant. The bones have been dated at 40 million years. The apparent separation between hominoids (the apes and man) and the other primates (the monkeys) occurred about 10 million years ago. Now the discovery of an ape skull at least 26 million years old, in those same Egyptian hunting grounds, sets another record.

Serious attempts to find the earliest man have been going on since the 1880s when a Dutch medical officer found bones in Java that he named Pithecanthropus erectus, or "the erect ape man." In 1927 similar remains of about three dozen individuals were found in a cave near Peking and, from the evidence of their braincases, it seems these people were eaten by cannibals. Both Java Man and Peking Man, now called Homo erectus, lived in the lower Paleolithic, a bit less than 1 million years ago. Then in 1959 Dr. Louis B. Leakey and his wife, digging in the Olduvai Gorge in Tanzania, found skeletons who were clearly tool users and lived 1,750,000 years ago. The Leakeys named him Homo habilus, a man who is able, mentally alert, and vigorous. Homo habilis walked upright, had hands like a modern human, could talk, was omnivorous, and about four feet tall.

Most recently scholars from Harvard discovered in Kenya, near Lake Rudolf, the jawbone of a hominid woman they dated, by radioactive means and from fossil elephants in the same place, as five and one-half million years old! So much antiquity surely places the Garden of Eden in Africa. Exactly when human types reached Europe will never be known but surely the Ice Ages had begun and the migrants had learned how to endure them. A man we would recognize as our own type, a member of Homo sapiens, lived in Europe 40,000 years ago when the ice was in an advance. Homo sapiens crossed the Bering Strait into North America (another date that keeps receding) at least 26,000 years ago and advanced south along a corridor in western Canada that was temporarily not covered by ice.

The question arises as to whether Homo sapiens, spreading all over the world, was a true species? Could, or did, Homo sapiens breed with the more primitive Neanderthals? Or were the others simply done away with?

Julian Huxley, the eminent biologist, had some interesting remarks about this new breed. He noted that man has eliminated all close competitors. He is the first organism to be represented by a single world-wide species. In other migrations, the wanderers undergo adaptive radiation and, like the finches in the Galapagos Islands, evolve into such specialized forms that they no longer breed together. But man migrates as an individual and in groups quite at will. His brain permits (and even stimulates) him to cross-mate very easily in spite of differences in color, appearance, or behavior; differences that would inhibit creatures that act mostly by instinct.

This single species which became so dominant had much to contend with during its Pliocene-Pleistocene career. Recent thought is that the Ice Age (a term that used to be synonymous with Pleistocene) did not begin just 600,000 years ago but more on the order of 3 million years ago and it did not come in only four waves but probably in eight. At intervals, ice caps 10,000 feet thick covered one third of the land area of the globe, most of it in the northern hemisphere but also affecting the mountains of South America, Africa (not far from the fossil finds of men more than 5 million years old) Asia, and Antarctica. The ice had profound effects on the land; its refuse created Cape Cod, Long Island, and the rocky fields of New England. The withdrawal of so much water into the form of ice lowered the sea level hundreds of feet making great changes on ocean coastlines. During periods of low sea level, land bridges appeared between Britain and Europe, Siberia and Alaska, Indonesia and Asia, New Guinea and Australia, with the natural result of numerous animal migrations. (Although the Bering Strait had been above sea level through most Cenozoic history, it flooded for several million years during the Miocene.)

Such drastic events as the ice invasions cause scientists to look for causes and the question is more than academic. Considering all the ice in the Antarctic and Greenland, it may be considered that we still live in the Pleistocene epoch, in danger of the ice caps melting and raising sea level around the world by about 300 feet. Many ideas have been suggested to explain the Pleistocene but none have found general acceptance. The world's two greatest ice sheets are under constant study to find if they are growing or receding although no one has suggested what might be done in either case.

During the Pliocene-Pleistocene many extinctions occurred, not necessarily all due to the deteriorating climate. When the ice finally receded, mammoths, rhinoceroses, and lions were gone from Europe. In North America rhinos, mammoths, mastodons, saber-toothed tigers, camels, horses, and many other animals were extinct as were giant sloths and glyptodont, an armadillo-type that had moved north from South America. Curiously some horses, elephants, and camels remained in North America until 8000 years ago, quite a long while after the ice was gone. Homo sapiens often is accused of wiping out many species and this has plausibility. Natural causes, the broad shifts of climate, could have reduced the population of quite a number of types and Pleistocene man have done the rest. Recent discoveries show that 8000 years ago man in North America had mastered the art of hunting by driving entire herds over the edge of cliffs to their destruction. Wasteful hunting has gone on in recent times as well.

The effects of the Pleistocene remain evident today in a great number of ways. The ice caps left the Great Lakes behind and determined the modern courses of the Ohio, Missouri, and Mississippi rivers. Northern Europe lost many native plants that could not retreat before the ice because they could not pass the barrier of the Alps. In Europe they never returned but the same kinds of plants, overwhelmed in New England, could move south because no great mountains stopped their progress. These were magnolias, sassafras, sweet gums, and tulip trees and they all thrive in North Carolina today.

Bird migrations, too, may be partially a result of the Pleistocene ice though this is hard to show except in some special cases. One recently reported concerns the bar-headed geese of the Himalaya Mountains. This is the observation of Lawrence W. Swan who could not understand why, or how, these geese would fly at an altitude of 30,000 feet from India to Tibet and make honking sounds up where they could hardly breathe.

Presumably few birds in the world migrated from nesting to feeding ground while a warm, equable climate existed almost everywhere. Migration would come as a response to changing seasons and to the search for regions where there was adequate food. Thus the bar-headed geese flew between India and Tibet beginning more than a million years ago when the mountains were still very low. At that time the monsoon rains could reach Tibet and the lakes and rivers were full. Then the earth began to move in the Pliocene and Pleistocene and the mountains caused by the collision of India and Asia began to grow very rapidly, reaching their present height in perhaps only a half million years. Tibet fell into a rain shadow and became increasingly arid but the birds persisted in their migration. The mountains continued to build, which certainly must have had an effect on world weather. The birds continued their time-honored course.

Swan understood this but could not understand why they flew over the highest peaks, such as Makalu and Everest, when they might have at least chosen the valleys at lower altitudes. His decision at last was that the geese keep their altitude because they navigate by identifying the tops of the Himalaya peaks and, furthermore, flying in gorges and ravines at night, when they usually travel, would be most dangerous. The birds may also recognize as dangerous the plumes of snow that sometimes blow for miles as jet streams sweep by and keep high to see and avoid them. As for honking, the flying geese may use this as an echo sounder to determine their distance above the peaks at night. The geology that makes mountains may often be overcome by the determination of living things.

". . . the greater size, strength, courage, pugnacity, and energy of man, in comparison with woman, were acquired during primeval times chiefly through the contests of rival males for the possession of the females. The greater intellectual vigor and power of invention in man is probably due to natural selection, combined with the inherited effects of habit, for the most able men will have succeeded best in defending and providing for themselves and for their wives and offspring . . . it appears that our male ape-like progenitors acquired their beards as an ornament to charm or excite the opposite sex, and transmitted them only to their male offspring. The females apparently had their bodies denuded of hair, also as a sexual ornament. . .

The races of man differ from each other and from their nearest allies in certain characters which are of no service to them in their daily habits of life, and which it is extremely probable would have been modified through sexual selection. (Shape of the face, squarenss of cheek bones, color of the skin, the nose, length of hair, presence of a great beard and so on.) These and other such points could hardly fail to be slowly and gradually exaggerated . . . of all the causes which have led to the differences in external appearances between the races of men, sexual selection has been the most efficient."
—Charles Darwin, <u>The Descent of Man.</u>

11

Sir Charles Lyell, one of the great men of geology and a contemporary of Darwin, accepted the idea of evolution but found difficulty in accepting Darwin's theory that it was all due to natural selection. He objected to assigning natural selection "more work than it can do." He saw how natural selection could eliminate the unfit but not how it could create the fit.

Darwin himself did not understand the mechanism by which species change and the science of heredity, genetics, did not really get a start until this century. Most simply the idea of genetics is that the individual receives his form from the union of the mother's egg cell and the father's spermatazoa. Within these are the genes which contain the messages on how the individual is to develop. Children from the same parents will never be exactly like one another, except for identical twins, because the genes received from the two parents never combine in exactly the same way. The offspring will be similar, however, within predictable limits. What brings about changes in this neat arithmetic are the events called mutations, without which there would be no evolution. Mutations or alterations in the genes occur as the result of high energy radiations (naturally occurring, as well as X-rays), and other causes not so well understood but presumably chemical and thermal in

117

nature. Each sex cell has thousands of genes, all of them offering the possibility of mutation. Some mutations are good, some make no difference one way or another, while others may be fatal. People sometimes carry recessive genes within them that do no harm to them but may be very damaging to their children.

Theodore Dobzhansky, world-famous geneticist, calculates that at the least one out of every five persons born carries a gene not inherited from either parent. "Mutation is not a rare phenomenon." Another way to say this is that 98 percent of the species that once existed have since vanished from the earth. With the calculation that there are at present 1,100,000 species of animal on earth, mutations have helped produce 100 million different animal species since life began.

Most genes reside unsuspected in the body. They may pass through generations without anyone knowing they exist. Then some challenge from the environment or a competing species may convert what had been a meaningless change into a positive benefit. For instance, there are regions in England where the moths were once light-colored, speckled, and banded in such a camouflage that they manage to be almost invisible against a tree trunk, thus eluding predatory birds. But in industrial areas where everything over the years has become black with soot, the moths have also become black. The simple mutation that made some moths black also made them less subject to being eaten by birds and thus they survived to produce more and more black moths in future generations.

How could one possibly guess the number of mutations it has taken to produce modern man from his ancestors in Cambrian times? That is, of course, an absurd question but we can look at the interesting problem of the way our bodies are formed by examining the individual features. The human body has been called a museum, a living witness to the past history of the species.

Man is one of the larger animals (though this may seem surprising for a moment) and seems to be becoming even more so. Prehistoric men were not much over five feet and, judging by the coats of armor seen in museums, the average man is 3 or 4 inches taller now than he was in medieval times. Size confers defensive and offensive capability and increased thermal efficiency.

The upright position acquired by man has made him a better hunter but at the cost of backache for so many modern sufferers. Though man has been standing up for at least two million years, one supposes, he still has not developed the ideal backbone for this posture.

A curiosity about humans, symmetrical in most aspects, is the fact of being right-handed or left-handed. This seems to confer no advantage. One might think, in fact, that it would be preferable to have both sides of the body equally strong but this is never the case. Why this condition exists has been studied but never explained. It seems that even in ancient man more people were right-handed than left-handed.

Among mammals only the armadillo still has armor. Man's skin, however, is actually an efficient protecting device, a means of sensory perception, and a heat control mechanism. Curiously, man can sweat to cool off while the higher apes cannot. Horses have the ability while dogs do not. No one seems sure why this should be.

While the breasts of most mammals are on the abdomen, they are on the chest in the human species and other primates. A thought about this is that it has some relation to living in trees, with the infant clinging on to the mother's chest while traveling. Why the human male should have nipples can only be explained as a vestige that natural selection has not done away with.

The pattern of hair on the human body seems to be mostly a matter of conjecture. Perhaps it remained on the top of the head as a protection from the sun and under

119

the arms and in the pubic region as protection in areas where the skin rubs together. Beards, Charles Darwin thought, are a cultural matter and consciously selected in some societies as a male distinction, something like the tail of a peacock. Some races have hardly any beards and apes have none at all.

Considering peculiarities, one has to wonder at the fact that no two people in the world have the same fingerprints. This has no evident virtue and presumably did not evolve to assist the police in finding criminals or maternity hospitals in identifying infants.

Unique to man is the development of the gluteus maximus, the buttocks, muscles which are used for walking upright. A racial aspect of this is the extravagant development of the buttocks among females of the Bushman tribe in southern Africa. Probably there is deliberate sexual selection involved but it has been suggested that this is also a great reservoir of fat stored up against the serious droughts that often afflict the Bushman territory.

Man glories in that most special of all things, his particular brain. He is the only animal who can think. When he comes upon something new, he does not necessarily respond in any predetermined manner. He can ponder it, imagine the consequences of this or that reaction, anticipate a mistake, avoid it, all without having to move at all. The human brain can imagine, calculate, predict. Even chimpanzees cannot plan far ahead, cannot control their feelings.

How this vast difference came about seems beyond knowing. Man has always been so conscious of his consciousness, his *knowing that he knows.* That, no doubt, is why he has imagined that a creator fashioned him quite specially. The discovery of his many affinities with animals lacking his special brain has often been rejected as implausible or downright insulting. That is why the idea of evolution is still rejected in some quarters. People can delude themselves into thinking something is not true simply because they do not like it. Being a sapiens means, among other things, being able to delude oneself. No one can imagine a dog or an ape deluding themselves. They may be deceived but they do not practice deception.

No more could any animal say, with Descartes, "I think, therefore I am." The problem of understanding our minds is like looking into a mirror. We can only get out what we put in. A machine examining itself. Some people call this human quality their *soul*. Perhaps we all know what they mean but just where *is* it?

One way by which paleontologists judge fossil animals is by the size of their brain. When it comes to the question of determining that the human stage has arrived, a large brain is a most important criterion. The brains of elephants and porpoises are larger than those of man but do not seem so complicated. The human brain is actually three brains that evolved one after the other. Earliest of these was the reptilean, which is programmed to act on instinct. This brain controls activities such as establishing a territory, mating, homing, and so on. Some scientists believe that man's capacity for violence lies here and also his capacity for obedience and following precedent. The reptilean brain would be the one that enjoyed ceremonial rituals and the one that desperately hangs on to prejudices.

The second brain to evolve has been called the old mammalian. This brain is the one that distinguishes what is real and true and concerns itself with the survival both of the individual and the species. The third brain, the new mammalian, is most highly developed in humans. It does the reading, writing, and arithmetic, writes philosophy and conceives of matters like nuclear fission.

Studies of brain waves show that these three brains sometimes act independently of each other. Such behavior leads to contradictions and momentary schizophrenia. It is almost a classic problem to say that reason leads you to do one thing while your emotions demand another. Such conflicts can result in mental illness. Animals with lesser brains presumably do not have such problems. Evolution of a large brain leads to success and survival for the species but for the individual it can produce anxious times.

Man inherited his sense organs from the past but none of them developed so well as his cerebrum. A dog's sense of smell and his hearing are much more acute than man's. The eyesight of an eagle is far better. We have no sonar such as porpoises and bats developed, nor can we sense magnetism like homing pigeons. To overcome these defects we have had to use our best legacy, brains, to develop instruments capable of telling us things other animals can find out without any equipment. (An interesting human inheritance are the bones that make up the hearing system. Very economically nature took the gills of primitive fishes, no use to land animals, and changed them to perform the new function of responding to vibrations in the air.)

That evolution did not make man perfect can be seen in the matter of his teeth. Children have to go through the awkward stage at the age of six or seven when they lose their milk teeth and for a while must go around with gaping holes in the front of their mouth. The biological reason for this is that at birth their jaws are too small to hold a complete, adult-size set of thirty-two teeth. Perhaps sexual selection has something to do with it, perhaps a smaller jaw was attractive to a potential mate, but somehow the human jaw seems to be wrong. Early mammals had forty-four teeth and the number has been reduced, probably because they were not all necessary for eating when hands could do some of the work in reducing the food. Yet even the present number seems too many when we produce wisdom teeth around the age of twenty-one that are ordinarily useless and sometimes do not even grow through the gums but become impacted instead. Such relic teeth often require very unpleasant operations in order to remove them. In experiments with teeth, as has been mentioned, sharks took the perfect direction. Should they lose a tooth, another will grow in its place within seven or eight days (no matter how old the animal).

Another somewhat defective inheritance in man comes from the problem of swallowing food for the stomach while breathing air into the lungs. Both must go through the part of the throat called the pharynx. Ordinarily a flap at the top closes for a moment while food is passing but sometimes we "swallow something down the wrong way" and this can be quite disturbing.

A not often appreciated structure is the palate at the roof of the mouth. As reptiles began developing into mammals, this bone appeared that separated the intake of air and food so they only had to meet once, at the pharynx. Reptiles do not need to breathe all the time and so their mouth can be filled with food for quite a period without any harm being done. For a mammal to stop breathing even for a few minutes would be fatal. Only the simple palate allows us to eat and still stay alive.

A dubious inheritance are the tonsils and adenoids which lie in the throat. In theory their design is to catch air-born germs but in practice they often become infected, in children, and swell so that breathing becomes difficult.

Another curiosity is the fact that in many mammals the same organ is used to carry urine and spermatazoa. The historical explanation is that, in the development of testes, part of the kidney system was borrowed to use for conducting tubes.

As remarkable as any human fact is that chemically we are like all other vertebrates. Our bodies are mostly water, mixed with salts, and these proportions are almost exactly the same as sea water. After a billion years, we still have not gotten away from the sea. It is literally in our blood.

The idea that humans evolved from less complicated forms seems particularly evident in the matter of reproduction. The old saying in biology class that "ontogeny recapitulates phylogeny," has a nice ring to it and means that the development of the embryo shows the stages that man has gone through since his beginning as two single cells. This is actually true only in a general way. First of all, the embryo repeats only the natal stages, not the adult ones, and it also skips along rather fast since it has quite a few hundred million years to cover. Yet exhibits of the human embryo at various stages do show the idea to be generally correct.

In the embryo the skeleton at first is all cartilage, as it was in the most primitive animals, and only later becomes bone. Not completely, however, for cartilage is left at the ends of the bones and continues to form there while bone continually replaces it during growth. When growth is complete, the cartilage is all absorbed and the shaft and ends of the bone fuse. Growth has taken place without disturbing the joints. (The tip of the nose remains cartilage throughout life. Is this because the nose is a feature particularly exposed to damage?)

It is remarked that every bone in the human head can be traced back to fish ancestors in Devonian times. Then it developed as a defense against the eurypterids, the primitive crabs that were challenging their living space. The human body in general can be compared almost exactly, bone for bone, with the first amphibians. The changes have been in proportions, not in the development of new bones. The human body is not a masterpiece of engineering design. It had no blueprint. It is the result of all the apparent accidents of mutation and natural selection. Considering its hit or miss nature, it works remarkably well.

London. November 4, 1970. (United Press International).

Chi-Chi, a female giant panda at the London Zoo that failed to mate with a panda from Moscow, has become something of a problem to her keepers and animal experts.

Chi-Chi has apparently taken a fancy to the occupant of an adjacent cage by the name of Paul. She chatters to him and pushes her bamboo shoots into his cage.

The problem is that Paul is not a panda but an onager, an Indian breed of wild ass that, unlike the panda, does not sit or stand on his hind legs.

"We are puzzled by the strange relationship," a zoo spokesman said today.

12

*A*nthropologists *say that in* some primitive tribes no one realizes that sexual activity has anything to do with the production of babies. While this appears unlikely, it may not be impossible. A hiatus of nine months does ordinarily occur between the two events. In the societies to which we are more accustomed, however, even small children are curious about their origin and want to know where babies come from. As a hangover from the niceties of the Victorian world, many parents find it difficult to tell them. It is known that a certain college girl at the age of twenty in 1911 thought she would become pregnant because she had to sit on a boy's lap during a sleigh ride.

Today everyone speaks more openly about the importance of sex but they are talking really just about intercourse and sex is much more important than merely that.

To repeat, it is the main cause for the origin of the species. In every species there exists a pool of hundreds of thousands of genes. As a result of sexual reproduction, there is a constant shuffling of genetic combinations. In a large population this has a conservative value. No dangerous genes can completely take over. Their effect will be muffled in all the variety. The human gene pool is so largely unknown in any detail that no one ought even to consider breeding for a perfect population. What might be the optimum for modern man today might not be that at all under different circumstances. With the enormous gene pool available in a large population, some humans should be able to survive even if things change drastically. If natural selection is not allowed to operate freely, man could paint himself into an evolutionary corner.

Such corners happen quite accidentally through force of circumstance and the results are clearly quite bad. This phenomenon has the name of genetic drift. It is the same as the island effect. In an investigation of several villages high and remote in the Italian Alps, the population was studied with the genetic problem in mind. The few families in the villages were all related. People had hardly moved either in or out for generations. As a result of this isolation there was a very high incidence of albinos, deaf mutes, and of mental deficiency. The matings called upon such a very small gene pool that harmful genes managed to overwhelm others. This is exactly the reason that most societies, even before they had any idea of genetics as a science, had strict taboos against incest. The offspring of any closely related couple usually had defects and weaknesses, hidden traits in the family line that were uncovered when two, harmful recessive genes joined together.

Something like 1500 diseases are known to be hereditary. One recently receiving a good deal of publicity is sickle cell anemia, a disease of Negroes from Africa that can often be fatal. It arose as an evolution in which the blood counters malaria and so, when malaria was prevalent, it had a function. Many people do not realize they have it because it may not be brought on until the victim is exposed to a lack of oxygen at high altitudes. It can be detected with a test using a single drop of blood and the afflicted are told to avoid heights. Positive remedies are being sought to control the anemia developed to fight malaria.

Among other diseases that are sexually transmitted are

the Rh factor disability, cystic fibrosis, diabetes, harelip, hemophilia, mongolism, and such things as fructose intolerance, in which the sufferer cannot eat sweets, fruits, or certain vegetables. There is a sort of race memory about all this and most normal mothers worry whether or not she and her mate are carrying recessive genes which, if joined, would bring out one of these ailments. For those with real cause to worry, there are doctors who specialize in human genetics to go to for advice. There are about thirty genetics centers around the United States.

Man is the most sexually active of all creatures and worry about his issue seldom bothers him when aroused. While birds and lower mammals have only one mating season a year, men may mate on any day. Human females are fertile an average of once a month and thus evolution gives them a twelve-fold advantage for reproductive success. Higher apes also have this monthly cycle and one may wonder if there is any real relation to phases of the moon, to the tides, or if this is a mere coincidence.

The reproductive advantage of a monthly cycle is more apparent than real, however, for it takes the human such a very long time to become sexually mature. A fruit fly can produce young when only a few weeks old. The infant fly needs no post-natal care, has nothing to learn, and would not recognize his parents. But human beings cannot even feed themselves for the first year and even to seem stupid they must learn a tremendous amount. Thus sex, among its many results, is the cause of human families. The utterly dependent infant needs a vast amount of help if he is to survive and, in turn, reproduce. Male parents, eagerly or reluctantly, must assist. But the male's role is marginal and so, out of boredom or, more likely, as a member of a hunting group, he has created societies, governments, economies, in order to sustain the species. Therefore it seems safe to say that sex is the basis of all civilization.

It may be reasonable to wonder at what epoch sexual

activity began to be pleasurable to humans. Evidently the higher primates find it so as well, but among the female mammals one is most likely to observe, the onset of sexual receptivity is a most painful matter. Bluntly, a cat in heat is clearly in agony. Some time during the long span of natural selection our ancestors found the nerve endings that would change mating from a terrible urge for release into an activity that would be eagerly sought out. Evolution is full of marvels but certainly there is nothing more marvelous than this discovery. Surely it has a great deal to do with the victory of mankind.

Evolution Stirs Coast School Debate. Special to <u>New York</u> <u>Times.</u>

Los Angeles, October 11, 1970. The teaching of evolution has become an issue in the public schools of California. Several members of the State Board of Education objected to guidelines for high school science courses that presented evolution as a fact, not as a theory, and made no mention of the belief that the universe was created by God.

"Evolution should not be accepted as a fact without alluding to creationism, which is felt to be sound by many scientists," said Dr. John Ford, a physician.

State Superintendent of Education, Max Rafferty, said the proposal would be rewritten to include mention of the theory of "creationism"—the concept "life was created in a short space of time rather than over a longer period through evolution and natural selection."

Dr. Rafferty, an outspoken conservative who ran unsuccessfully for the Senate last year, also stirred up some controversy with a letter he wrote to local school officials urging them to inspect the lockers of high school students suspected of hiding drugs.

13

*N*ow *man himself has become* a very conscious evolutionary force. By design and through selective breeding he bring about changes thought desirable for himself in a great variety of animals and plants that he manages to control. Ants enslave other ants and many kinds of animals live in relationships with different species that quite obviously are useful. No creature until man, however, emerged as so clearly superior that he could deliberately command so much of nature at his bidding.

This domestication practiced by man probably did not begin as an intellectual matter. People generally believe that man's first domestic relationship started with the dog, between 40,000 and 60,000 years ago. No one can know just how this old friendship began but Konrad Lorenz has written a charming, educated guess about it in his book, *Man Meets Dog*. The Austrian scientist visualizes a Stone Age scene where a small band of men and women struggle across a plain in the dark of night. They have been driven from their own land by a more powerful tribe. Often in their flight they stop and listen, out of fear not only of men but tigers and other animals. On their own grounds they had no such fear because wild jackals usually hovered around, waiting for refuse and setting up howls when they heard an intruder. Now no jackals with their keen hearing and sense of smell were keeping watch. But the next night around their camp fire the little band knew that jackals had found them and were waiting in the dark. The next day as the people set out again, the leader deliberately left behind some meat on the trail, so the jackals would follow. When the small band settled down again permanently, the jackals were with them and over the ages they became bolder and tamer. They would go hunting with the men in hopes of getting some scraps from the bill.

One day the hunters wounded some animal, perhaps a wild mare, but she managed to get away, leaving a trail of blood. To elude her pursuers, she doubled back on her tracks but the men did not notice. The dogs did and followed her scent, so obvious to them. Now the hunters noticed the jackals were not following and they went to where they heard the animals barking. They finished the kill and the leading hunter threw some of the mare's guts to the dogs. One of the dogs, while eating, looked up at the man and shyly wagged his tail.

The skulls of Spitz-like dogs, clearly domestic, have been found in the remains of dwellings on the shores of the Baltic dated about 20,000 years ago. These skulls are only found associated with human remains. Lorenz sees a little girl out somewhere at play, hearing a whimper and coming upon a jackal lair with just one little pup in it. The mother and the rest of the litter were dragged off by a tiger. Womanly instinct makes her pick up the baby, bring it home and give it food. When the father discovers this animal in his house, he wants to kill it but the little girl's tears stop him. As the baby animal grows, the animal learns who is the master and, obedient to his instinct for hierarchy, begins to follow him. At the close of the story we see the little girl watch wistfully as the man and his animal go off to the hunt.

At some time these dogs in the north began to mate with wolves. Obviously today many dogs have wolf blood and a cross between the two will take, producing healthy, fertile offspring. Huskies, Samoyeds, and Chows all come from the north and all have obvious affinities with wolves, not only physically but in terms of behavior. Behavior is highly resistant to evolutionary change and Darwin remarks about a dog who had one wolf as a great-grandparent but was in all but one way completely civilized. The dog would approach his master in the sideways subservient motion wolves assume when coming up to their leader.

Dogs retain some but not all the patterns of wolflike activity. They howl with the same voice but, fortunately, less often. Wolves may bark and wag their tails but dogs do this frequently. Wolves have a variety of very clear facial expressions. Dogs have the same ones but not in such a highly developed form. There is a strong social hierarchy among wolves and the top animal has the first choice of food and mate. Among wolves, male and female are at least partially monogamous and the father brings food to the lair while the female is suckling the pups. Male wolves seem to love the pups and will submit to all sorts of foolishness when the puppies play with them. They look on indulgently while the youngsters chase their tails, engage in mock battles, and attack the adults in a way that would surely not be tolerated from another adult.

Wolves usually hunt in packs and the acknowledged leader directs the attack. When wolves howl it is not necessarily a sad cry at all but a complicated system of communication. The howl is much like a song, with many different notes, and each wolf has a distinct voice recognized by all the others. Wolves are very strict about their territory and carefully mark its boundaries by urinating on markers such as tree stumps. Any strange wolf foolish enough to venture into this marked-off region is warned off by barking. If he persists, there will be snarling, a display of teeth, ears turned back, and a bristling mane. If this is not sufficient, there will be a fight.

In a fight between wolves, neither will be killed because

they have a strict code of honor. The victor must stop the attack if the loser presents his rear to him or lies down on his back to expose his vulnerable neck and stomach. Having acknowledged his own defeat, the loser is allowed to get up and run away with his tail between his legs.

Much of this can be seen in the behavior of dogs, of course. Although restrained by fences, leashes, or automobile windows, they go through all the same bluffing, barking, marking on lamp posts, the same effort to find out who is the tougher. (And one of the things guaranteed to infuriate a dog is for him to be on a leash and see another dog running free.) Dogs are very aware of each other as individuals and anyone who has had a dog in neighborhoods where they have some freedom knows about the feuds that exist between certain dogs, the houses they bristle at, the particular posts and trees they insist on marking.

When it comes to young dogs and females, domesticated males retain the wolves' code of honor. They will suffer all sorts of indignities from a playful puppy without harming him. Until he is about six months old, a young dog is untouchable to other dogs. Then he must learn to be subordinate until strong enough to assert his rank. A male dog will never harm a female one though he will pursue her relentlessly if her sexual scent excites him.

People become much more sentimental about dogs than about any other animal. Literature is full of stories about dogs who will not leave the master's grave, about dogs who risk their lives to save children, about their interfering when members of a family fight, of defending their masters and children from attack. All these things and more are true and all come from the heritage of the wolf pack. To dogs, their owner is not just a pack leader, but a "superleader."

"Man's best friend," the "faithful dog" often seems so close to understanding that people try to teach them to speak. (All they succeed in getting is a bark.) They do appear to understand when people are unhappy and will try to comfort anyone crying. They fear people who are intoxicated. Dogs understand tones of voice very well, whether tender or stern, and they soon learn words like "walk," "go out," and "dinner" so that people spell them when they do not want the dog to know what they are talking about.

How man came to teach dogs can only be a conjecture. Darwin thought that certainly no one would think of teaching a dog to point but, when a hunter observed a dog doing this, he could see by his stance and his nose that he was scenting a prey. Such dogs would be valued and more attention paid to their breeding. Perhaps dogs learn well because, in their long association with man, they have learned to live less by instinct and more by feeling. Man considers a good dog one that is anxious to please and this is the one who will obey and try hard to understand instruction. A cringing, too subservient dog is not attractive, however. Most people like a measure of independence, along with obedience. Dogs are deliberately bred to enhance particular characters.

Breeding dogs to preserve or improve various physical features is easier than selecting their mates for disposition and such eugenics have gone on for many centuries. Greyhounds and beagles are both chosen to hunt rabbits but with greyhounds the object is more speed while the beagle is kept short, so he can dig in burrows. Dachshunds and Sealyhams have also been selected as diggers. Among the great number of recognized breeds most of them, including poodles, have been specialized for some kind of hunting. (This applies even to the St. Bernard used for rescue work but which must hunt to find humans lost in the mountain snows.) Dogs excel at hunting, naturally, because of their fine noses and ears. Yet they are so adaptable that they learn to do police work, control and guard sheep, lead the blind, and even guard the beaches against infiltrators during wartime. The control of evolution in dogs has been such a success that their survival as a species remains assured as long as man is dominant.*

As an evolutionary force, man has changed matters much more rapidly than could happen if selection were only natural. A large variety of dog breeds have now been created—116 breeds are recognized by the American Kennel Club. But when he began to domesticate plants he truly showed what might be done to nature. Now there are new breeds of roses, new varieties, offered every single year.

Man began to be a selector of plants, a farmer, about 10,000 years ago according to carbon dating. At that time the world population was about five million people and the last great glacier of the Ice Age was retreating to the north. Deliberate planting of crops probably began around the Tigris and Euphrates Valley in the Near East, which was then fertile. The original wheat was a variety called emmer which proliferated beyond its original abundance simply because man planted it and disposed of invading weeds by cultivation. But the emmer wheat cross-pollinated, not because of any help from man, with a noxious weed called goat grass. This cross produced a new variety called bread wheat, now one of the most important foods in the world. The early farmers took advantage of this new wheat and saw to it that it spread to many lands in a way and at a speed that would never happen by natural means. In the next millennia men found they could also cultivate plants we know as peas, lentils, barley (for beer), linseed (for oil), and grapes (for wine). Under cultivation, however, some plants were altered. An ear of wild emmer wheat, when ripe, breaks into separate parts, each with one grain protected by chaff, and covered with a beard that will catch on to the coat of any animal that passes. After falling to

---

* One may wonder if dogs actually are a single species any more. Can one imagine breeding a Chihuahua to a St. Bernard or Husky? The offspring could hardly survive, even if mating succeeded. It seems that man has come close to creating two new species out of one.

the ground the grain immediately starts to germinate. Bread wheat and cultivated emmer wheat do not shatter upon ripening. These were the ones favored by farmers because the stalks can be cut and carried without breaking. Such grains have to be threshed before the grain is available to become flour.

In cultivation plants often lost their ability to cross-breed because those which pollinated themselves were the ones that gave higher yields. The crops were improved, in the eyes of the farmer, but natural selection by sexual means was eliminated. The growers of wheat did not know this but only enjoyed their greater prosperity.

The effect of successful farming on the human race is very obvious. Fields that would yield more than a subsistence for one family, grain that could be stored, meant that cities and civilizations were possible. Man's interference with evolution made everything else perhaps inevitable.

What happened when man took over nature's role can be seen in the New World as well as the Fertile Crescent of the Near East. Here the discovery was the seed of a wild grass that popped when exposed to heat. This was primitive corn, and researchers believe man discovered its edibility by accidentally heating it. Thus one may visualize Indians, perhaps in Peru 9000 years ago, having the world's first feast of popcorn. Christopher Columbus discovered corn on November 5, 1492, shortly after arriving in Cuba. The great explorer had sent two of his men to scout the interior of the new land and they returned to tell of the Indians' food, "a sort of grain they call maize which tasted well baked, dried, and made into flour." Later discoverers found that wherever they went in North and South America the Indians used corn as their basic diet. It was their grain that could be stored in surplus, thus permitting men to congregate in towns and cities rather than farm just to keep alive. A ruling class could hoard it in great quantities and use their possession as a source of power.

Its use everywhere in the Americas means that it must have quite an ancient history. Originally the wild plant could scatter its seeds itself, in the way of other grains, but its first users naturally chose those plants which had more seeds on each cob, thus beginning the selection toward the modern corncob which has several hundred kernels. One researcher suggests that the original corn had a low survival value, a limited range and was well on the way to extinction when man first began to use it. The Indians, learning its virtues as a food, dispersed it over two continents. The once insignificant plant became the dominant crop over a vast region. Hybrids would first occur naturally when corn seeds became fertilized by adjacent wild relatives but in time man began to create hybrids himself. Thus, by dispersing the plant and breeding it, the human race once more became an evolutionary agent.

Modern corn could not exist at all without human intervention. When an ear of corn drops to the ground, numerous seedlings emerge and all of them engage in such wild competition that all of them die long before they could reach the stage of reproduction. Corn would be extinct if man did not disperse its seeds for it.

Another example of plant migration at the hands of man is the sweet potato which began either in Mexico or Peru. It had reached New Zealand even before the time of Columbus. The Maoris of New Zealand called the sweet potato "kumara," which is the name it had in Central America. Many centuries ago people from the Americas must have traveled westward across the Pacific, perhaps like Thor Heyerdahl on his raft, Kon-Tiki, and brought the sweet potato with them when they landed on the Polynesian Islands. From there settlers took it to New Zealand.

Now, of course, all sorts of plants have been carried far from their lands of origination. White potatoes from Bolivia or Peru, oranges from China, lemons from Italy, grapefruit from Malaysia, tomatoes from South America,* rye from Central Asia, all the various flowers and trees that have been exported—it is really quite staggering what man has done to the face of the earth just by planting alone. Now, too, man creates hybrids not only by cross-breeding but by exposing plant seeds to irridation and chemicals to create artificial mutations. Such procedures produce many harmful or useless variations but diligent observers note those mutations that are beneficial, nurture them carefully and produce new breeds at a considerable rate. Workers in Mexico produced a new kind of wheat that proved to grow extremely well in starving India, but the seeds were red. Indian farmers only knew about wheat grains that were amber in color so this red grain was indignantly rejected. Within three years the scientists produced the grain of exactly the same quality but the desired amber color, which the Indians accepted without question.

Man evidently began domesticating plants some thousands of years before he tamed any animals except the dog. From finds in archeology it is known that sheep and pigs began to be cared for about 7000 B.C.

One of the earliest examples of cattle domestication was found at Non Nok Tha in northeast Thailand, a Bronze Age settlement of about 5000 B.C. Bones of water buffalo and zebu, as well as two extremely rare species, bouteng and kouprey, were found among the remains of men who had been rice farmers. Horses and camels were not domesticated for another thousand years, just about the time that Bishop Ussher said the world began.

Chickens are types of pheasants whose ancestors roamed the jungles of India and Ceylon eastward to Indonesia and China. These fowl differ from other pheasants in having

---

* Grown in the United States only as a garden ornament until 1830 because it was thought to be poisonous.

a comb, wattles around the head, and a more arched, curving tail. Seafarers took them to the Philippines and Polynesia and they reached the New World before Columbus. Chickens came to Europe as a result of the Persian invasions of classical Greece and Aristophanes called them "the Persian bird."

The domestic cat first appears in history as a religious symbol in the temples of ancient Egypt. One of the goddesses of the time was Basd, who took the form of a lioness. The presumption is that the Egyptian priests preferred having a small version of a lioness, rather than the real thing, living among them. Cats were supposed to have been bred as palace guards in Siam and must have been consciously selected for the quality that Siamese cats still have. If permitted, most Siamese cats will jump up from behind a human and land on his left shoulder. Perhaps in the palace they went a step further and attacked his throat from this position. Angoras and Persian cats, now bred as ornaments, must have been originally tolerated because they kept away rats and mice. An idea is that cats took up residence with man when he began to store food in granaries. This certainly attracted rodents, which would in turn bring on the attention of cats. Presumably man did not decide to keep cats. The decision was made by the animals instead. Some authorities say that cats never have been domesticated; living with human beings is only an arrangement that suits their pleasure. Quite certainly a cat brought up in a house is much better able to take care of itself in the wild than a dog used to human care would ever be.

Domestication and breeding for special qualities means that a species will survive because man has found it valuable, but such care does not always produce desirable effects. The turkey raised for feasts is so stupid that it may drown while standing in the rain with its mouth open. Certainly wild turkeys never succumbed for such a cause. Until recently Collie dogs were bred to produce a fine, thin head considered to be very beautiful. This continued until it was realized little room was being left for the brain. In 1971 the thoroughbred Arabian race horse, Hoist The Flag, shattered the anklebone above his right hind hoof simply by running on a perfectly clear, smooth track. Thorough-breds have been selected for speed to such an extent that their thin legs are extraordinarily fragile. Hoist The Flag was considered the most promising horse of the year and he was not shot, as often happens to horses with broken legs, but given extraordinary surgery and care so he might be saved to serve as a very valuable stud.

Recently man's control of animals turned out to have another profound effect. Archeologists, from experience, believed they could tell the difference between wild and domestic bones simply by feeling them. No scientific basis for this thought existed, however, until workers at Columbia University began to examine, under powerful instruments, bones whose status was already known. They discovered that the crystal structure in the bones of wild animals was quite different from the structure in domestic animals. No one yet knows what brings about the changes but penned animals would lead quite different lives; be less active, perhaps overcrowded, and eat differently, either for better or worse.

It turns out that excessive breeding may sometimes be a calamity among plants. During the 1970 growing season a good portion of the corn crop in the United States suffered a very serious blight. Much of the American corn was being grown from the same highly successful hybrid seed. The blight was a mutant strain to which this hybrid was by no means immune. The lack of diversity, of any genes that would help the corn resist this attack, could in an extreme case have led to extinction. If all of the species were exactly alike, blight could have killed it all off. Inbred species are very vulnerable. Some of those concerned now work to keep a variety of ancient strains under cultivation so there will be a choice of genetic material to work with in future breeding. Domestication of plants and animals has made possible the life we lead but interference in the process of evolution is not in all ways beneficial.

The love play of animals is full of variations. Mammals press themselves together in their sexual excitement, snuggle up to each other and touch each other's bodies with nose, mouth and paws. They can nibble, bite, scratch and stroke, or tug at each other's coats and pull out hairs. Such activities can last minutes, hours, or even days before any attempt at copulation is made . . .

Strong males fight one another for desirable females and these disputes often end in the death of one of the contestants. Camel stallions bite and spit; the steppe antelope's antlers become a murderous weapon; baboons stretch up to their full height, open their mouths wide in fury, and poke their tongues out and in their jealousy even lazy sloths fight slow-motion duels with one another.

—Fritz-Martin Engel, <u>Life Around Us.</u>

14

*D*o *reasons exist for every-*
thing we can observe in nature? The human mind by its
nature assumes that everything has an explanation. Scientists live by asking "why" and, if they cannot find an answer,
they do not shrug and say it is not there. They still have not
found all the reasons. "We don't know all the answers yet.
We don't even know all the questions."

Yet many strange and marvelous things in nature have
been explained. All sorts of answers about the millions of
different ways things look and behave.

Why can all members of the cat family retract their long
claws? Claws, so useful for capturing prey, are very awkward
for running or even walking and so the ability to sheath
them has strong adaptive value.

Of what use is the snake's infrared detector that can accu-
rately measure heat to within .001 degree Centigrade over
a wide range? Snakes use this ability to find the location of
prey and to be warned of potential aggressors. It is their
most important sense perception.

What is the function of poison ivy? As a defense against
predators. This is a passive defense; part of the plant must
be eaten before the effect of the poison is felt. Animals
cannot afford to be partially eaten before the attacker

realizes its mistake so they get immediate attention by spraying, biting, or stinging.

The evolution of life has produced so many wonders that they defy classification. Where does a question belong that asks why some animals clean themselves while others do not? Why do dogs get so dirty while cats all clean themselves? The answer typifies all species; those animals that live the life of nomads lack any instinct of hygiene but any animal that remains in one dwelling place will instinctively be clean.

And this behavior pattern leads well enough to birds because most of them are fastidiously clean. (This does not apply to Adelie penguins.) Beyond the virtue of cleanliness, however, which recommends them to grandmothers, some kinds of birds behave in ways that are positively delinquent. Often funny, unless you are the victim, the most eminent rascal of all may be the magpie. (The author once had a brief but very delightful romance, in the Washington Zoo, with a magpie who came from southeast Asia.) Magpies are great thieves and will steal any bright object they can carry in their claws or beak. They are also chatterers, noisy birds, and so the saying goes about a talkative person that she is just like a magpie.

Magpies belong to the enormous family of birds known as Corvidae. Almost all of them are bold, noisy, aggressive, very active and conspicuous. A good number of them are robbers. As a group, they are very familiar to mankind and distinctive common names have been given to them that are household words in many languages. Among the one hundred species assigned to the Corvidae, along with magpies, are such well known birds as crows, jays, ravens, rooks, and jackdaws.

It has taken about two million species of birds since Mesozoic times to produce the present 8600 species. Why did the ancestors of the present hundred species of Corvids arise in Miocene days 25 million years ago? What set them on their vulgar, pushy ways. No one can tell from fossil remains but the competitive Corvid character, no matter how unpopular it might make them in a crowd, has obvious survival value. And you notice such creatures, even if you deplore their manners.

Corvids are best known in the Northern Hemisphere. They will eat almost anything they can swallow—animal, vegetable, or mineral. So they have been able to adapt themselves to changing conditions in a number of habitats. Some species, however, have curiously limited distribution and others have ranges that are widely separated by regions in which the species does not live.

They are thought to have begun in the Old World but jays reached Central and South America not long after. The crows appeared later and only settled as far south as Honduras. They are tougher than jays, however, and better fliers and have made their way to the West Indies, Philippines, Australia, and the smaller Pacific islands where jays are unknown. All the genus Corvus say "crow," "craw," "caw," or "krahe."

The behavior of crows is much too varied to be called just instinct. They play games like waking up a sleeping rabbit by pecking it on the head and do things like settle on the back of a drowsy cow and suddenly set up a great clamor, then fly off cawing happily. Crows in fields post sentinels and the sentinels can tell the difference between a man with a stick or one carrying a gun. Several crows will gang up on a dog to steal his food. They dunk dry crusts of bread in a stream to soften them. Finding a clam or an oyster, they will fly high with it, drop it on the rocks or a highway, and swoop down to eat the morsel inside.

Crows seem to have a strong social sense. If one is injured, other crows will gather around him. They will bring him food. They will try to lift him. In any catastrophe, a crow's call for help will always bring companions who will do what is possible to assist him. Without specialized eating habits, crows have managed well in a world so radically changed by man. They may eat sprouting grain but they devour enormous amounts of insects as well. They are believed to be the most intelligent of birds. It has been shown that they can count to three or four and learn to associate various noises and symbols with food. Among themselves they have a quite extensive language. Though they do learn to say a few words, they are not so adept at this as parrots or mynah birds.

Crows are constantly persecuted because hunters hate them for killing nesting fowl and game birds. During their winter roosting periods thousands are slaughtered by dynamiting. They nest in large flocks and the male feeds the female during incubation. From their European relatives, the rooks, comes the word "rookery" which now stands for any site where many birds congregate with their young.

Another Corvid is the chough of Europe, a noted aerialist who often tumbles about in the sky, apparently just for fun, and even turns somersaults while flying. The Corvid nutcracker specializes as a memory expert who stores great quantities of food, nuts, and remembers later where they are.

In the curious Corvid pattern of habitat, the beautiful azure magpie lives in eastern Asia and Japan and then appears again in the highlands of Spain. The jay scrub ranges from the state of Washington to Mexico and then is not seen again until Florida. Few corvids migrate. They are large, well-settled, long-lived birds who find no necessity in changing their homes.

But why are they so mischievous, such notorious thieves? Why do they like to pick up bright objects and fly away with them? One might suspect they want them to decorate their nests but usually they do not take their watches and jewels and car keys home. They simply play catch with them in the air, dropping them, and then swooping below to retrieve them, and then finally tiring of the game, let them fall where they may. One might surmise that what began as a nest building activity has evolved into sport for sport's sake. Or perhaps they have developed a vulgar sense of humor and learned to laugh at helpless humans from whom they have retrieved some special treasure. Do humans know if they are the only creatures who can laugh?

In human terms, birds obviously have no moral code. Many are shocking parasites. Herring gulls swarm around single ospreys and pelicans who have just caught a fish, and harass them so that they drop it. Bald eagles sometimes successfully pursue ospreys in the same way, with theft in mind. With jaegers and robber gulls, such behavior is their way of life. Skua birds in the Antarctic live right alongside the Adelie penguin rookery. They do not attack the adults but go after any egg or chick that is left unguarded.

An evolution that seems even more deplorable is that in which cowbirds and cuckoos lay their eggs in the nest of other species to let other birds do their hatching for them. (The honey guides and weaver finches of Africa also do not build nests or rear their own young.)

The various cowbirds of the world show how this practice may have worked out. Some of the cowbirds in Brazil and Argentina do build nests but only do so if they cannot find a nest abandoned by others. Others have learned to leave their eggs in nests where the rightful owners will hatch and raise them but these cowbirds have not yet completely lost their nesting instinct. During courtship the Shiny Cowbird picks up nesting material and starts to build a nest, but never finishes the job. The Shiny Cowbird has not yet learned the new mode very well. She wastes many eggs by simply laying them on the ground if she cannot find a nest and she also may put more eggs in a nest than the foster parents can handle. As many as thirty-seven Shiny Cowbird eggs have been found in the nest of one Ovenbird.

American cowbirds have traveled further along this murky road. They usually lay only one egg in a particular nest but it is the nest of a smaller bird, often an oriole, which has smaller eggs. The cowbird egg usually hatches a day before the other and the chick, in any case, is larger and demands more of the food brought by the foster parents. He does not attack his rivals in the nest but may often starve them.

Female cowbirds establish territories and watch other species of birds build their nest within them. She knows in advance where her eggs will go. Apparently the sight of nest building is the stimulus that causes her to ovulate. She will lay four to five eggs on successive mornings, each in a different nest during the few minutes the owner is away feeding. She does not remove any of the owners' eggs.

But some cuckoos (not the American variety) have improved upon this. The cuckoos everywhere say their own name and so they are called the same from Japan to western Europe. They are also the basis for the word "cuckold," one whose wife is unfaithful.

Eggs of the European cuckoo are all the same shape or size but the females specializes in laying eggs of the same color as her victim. Thus warblers' eggs are blue and so are the cuckoo eggs placed in its nest. The female has the same watching procedure as the cowbird, lays her egg just as swiftly, but after doing so takes out one of the other eggs with her bill and either swallows its contents or drops it to the ground. The willow warbler builds a roofed nest that is too small for a cuckoo to enter so the female lays her egg on the ground and flies with it up to the nest, carrying it between her claws.

When a naked cuckoo is hatched it immediately throws all other chicks or eggs out of the nest. Sometimes two cuckoo eggs have been laid in the nest by two different parents. If one hatches first it will immediately destroy the other egg. If both are born at the same time a battle will happen which neither may win. Should the two little cuckoos both survive for three or four days they forgot their antagonism. The cuckoo's instinct, its reproductive pattern,

152

is so complex, so refined that the young Shining Cuckoo of New Zealand, who has never seen his parents, is born knowing he must migrate 2000 miles to the Solomon Islands. He has the innate urge to winter there and somewhere in his little body there resides the race memory, the skill necessary to make the correct flight.

Mention of bird migration recalls those aspects of avianism which make them so interesting to mankind. City and suburban dwellers who generally see nothing much more than pigeons, starlings, or robins making short, jumpy little flights tend to forget just how magnificent the sight of flying birds can be. A hawk circling high in the sky, the pelican diving like a bullet into the sea, a flock of ducks taking off from a lake, sandpipers wheeling over a beach make men sometimes write poetry. Others have tried to understand flight mechanically to learn how, by artifice, to do something which birds can do without instruction. Young eagles are often reluctant to leave the nest and do so only because their parents, knowing it is time for them to fly, simply stop bringing food for their voracious stomachs. The mother will often be near the nest when the fledgling at last dares take wing and they have been seen supporting a young bird on its back if it falters for a moment. Then it feels the draft of strong air currents flowing up the steep side of the cliff and begins to soar on its own.

By taking the turn toward life in the air, birds made themselves more vulnerable on the ground but the advantages of their way of life are many and a vast number of birds thrive in the world. Those which nest in trees find safety there from predators. Many insect-eaters find their food in the bark of trees. Humming birds discover nectar in flowers above the ground. Some birds nest on the monuments and ledges of cities during the night and commute to the country to eat during the day. Flight opens up a great area in which to search for food. Flight puts birds beyond the power of weather because when it turns bad they can simply move to another state or another continent.

Such migrations must historically have been a response to a change in the seasons. Many migration patterns were formed during the ice of the Pleistocene epoch. Birds who live on insects would starve if they did not move when their food supply disappeared during the cold. The seed-eaters are much less likely to migrate because they have learned how to find food under all circumstances. Birds that live on fresh water must also flee before they are frozen into the ice. A dense bird population, too, may use up so much of the available supply of food that the birds must move on to a fresh territory.

How do birds know when to migrate? Controlled experiments show that when days are long, the birds spend more time in activity and less in sleep. Somehow this stimulates secretion of hormones and the sex organs begin to grow. In a laboratory, migrating types of birds were kept away from natural light and subjected to artificial situations with the days getting a little longer every evening. When these birds were freed they headed north, even though it was in the dead of winter. Their glandular chemistry gave them the message. This is breeding time. Temperature did not provide the message. Birds do pay some attention to the weather, however. Canada geese go north only as the temperature line of 35° Fahrenheit moves in that direction. They stay a few degrees above freezing.

Canada geese are one of the most familiar migrators in North America. Each October about three million of them head south along four traditional flyways. In recent years their habits of many centuries have been changing, however. Many were oriented to the Atlantic Flyway, some of them going all the way to Florida. Now a majority of these spend the winter on the Delmarva Pensinsula between Chesapeake Bay, Delaware Bay, and the Atlantic. A lake in North Carolina that once had 120,000 wintering Canada geese had only 24,000 in 1970. At a location in Florida a population had dwindled from 20,000 to 2000. The story is the same on the Mississippi, Great Plains, and Pacific Flyways.

Two reasons are given for the change of habits. Many states have created waterfowl sanctuaries, wetlands, where the geese are safe from hunters. Canada geese are smart and quickly learn where these refuges are. The other explanation for the new trend is that before the land was tilled the geese had to live on wild grains, not very high in nutritional value, but now the birds can find corn and other superior grains, which help them survive the freeze and snow. The situation seems good to them and they do not waste their energy in unnecessary flight to the south. Canada geese migrate in families and the older generation teaches the young birds the location of ancestral wintering grounds. That this knowledge is learned, rather than locked-in instinct, is shown by the fact that it only takes a few years for the birds to adopt a new habitat. Conservation men in the south have complained about this seduction of their geese by states farther north but it seems unlikely anything will be done about it.

(As another example of changing migration patterns, bird-watchers in Switzerland noted recently that swallows flying north in the spring have discovered the four-mile Grand St. Bernard Road tunnel from Italy to Switzerland and now take this shortcut rather than make their way over the 8000 feet Alpine peaks.)

From the evidence it would seem that geese navigate by eyesight and memory during their migrations. Many birds, however, navigate by night when landmarks may not be clear. Others fly over hundreds of miles of water where there are no landmarks at all. The world's champion navigator, the Arctic tern, who summers in the Arctic flies all the way to the Antarctic during the summer season there and he has to fly 11,000 miles, much of the way over water, in order to do this. In the Antarctic, experiments have been made in which blindfolded penguins have been flown hundreds of miles away from their rookery and then released. These flightless birds made their way home safely over great distances of ice and snow they had never seen before. Birds must have more than one means of navigation.

Tests have shown that many birds carry within them an imprinted chart of the stars and know intuitively how to steer by a particular configuration of stars that always exists during their time of migration. This may be how the swallows of Capistrano always manage to return on the same day in spring. The experiment in the Antarctic shows that penguins must use the sun in finding their way and know, by the angle of the sun over the horizon, that it is time to return to the nesting ground where they were born, after the six-months' night of winter at sea.

People have speculated that birds may also use the earth's magnetism to discover where they are. Experiments on homing pigeons seem to show that this is so. They do use the sun and familiar landmarks, but they can arrive at their destination without these aids. To show this, two groups of pigeons were used. They came from the same flock. One group had a magnetic bar glued to their back while the other had a plain brass bar of the same size and weight glued on. All the pigeons were taken in closed trucks to the same fairly distant location and then set free. The day was cloudy. The birds carrying brass bars soon made it home but those with magnets glued on, which completely upset the magnetic signals from the earth, had an extremely difficult time making the return trip.

The Canada goose is monogamous. He mates for life and remains forever faithful but, if his mate dies, he will seek a new female. This condition, this manner of living is supposed to occur most often in species where the young need considerable training after they are born. Yet monogamy does not obtain in all such cases. A kitten must be taught how to clean itself by the mother and, most importantly, how to hunt and catch mice, a matter of instruction in which the mother takes particular care. But the kitten who has to learn, seldom if ever knows its father.

The trouble with explanations of things like monogamy is that they are merely expressions of opinion, not matters of scientific truth. One may look at a kind of behavior and reason that such and such must be the reason but how can one experiment to find out why some birds mate for life, others just for the season, while with others the whole affair is over in a matter of minutes?

One such promiscuous creature is the fearless, pugnacious little hummingbird. The male hummingbird's courtship is spectacular as he swoops and arcs around the female, displaying his powers and his plumage. Mating often takes place in the air and then the male immediately departs, looking for other conquests. Hummingbirds are found only in the Americas and out of 319 species, in only one does the male help rear the young.

When speaking of the mating of birds, the actual subject is *everything* about them, their behavior and their appearance. Nothing about birds is unconnected with their sex life, the survival of their species. What attracts many people to them is what the birds themselves find very attractive. This is their beauty and usually it is the males who have evolved in beauty in order to attract females. Humans admire their plumage and their song. (Though Darwin points out that they are also "ornamented by all sorts of combs, wattles, protuberances, horns, air-distended sacs, topknots, and naked shafts" which we do not feel quite the same way about as birds do themselves.)

The trend toward gorgeousness in males occurs in almost all species, until the gaudiness of birds of paradise and peacocks seems almost an end in itself. The peacock is so vain that it will show off its feathers to poultry, or even to pigs. Such evolution can become self-defeating since it ignores another function of color, camouflage. To blend in with the scene, forest birds are most often green, ground birds mostly brownish, sea birds black and white. Birds in the sunlit foliage of the jungle are often multi-colored to blend with the surroundings. Ptarmigans of the north become white in winter. In general, color becomes brighter as birds become more tropical, as similarly happens among fish. For camouflage, most females are relatively dowdy, since they must remain tied to their nests for so long. Males moult in the fall and are comparatively inconspicuous until mating time returns. Color also serves as a means of birds finding their own species to mate with, an identification card so that time will not be wasted in useless courtship. Birds have a very acute sense of color; a pigeon in an experiment was able to distinguish between twenty different shades of red. Bird color can be influenced by diet. The flamingoes at Hialeah race track in Florida are fed a diet of shrimp to make their feathers a darker pink.

The beaks or bills of birds are all adaptations to the many different diets birds have specialized in, filling so many ecological niches. Birds who eat insects have thin beaks while seed-eaters have thick beaks. The carnivores such as eagles grow strong, hooked beaks. Hummingbird species have beaks adopted for eating in particular flowers. Skimmers have a long lower bill to pick up small aquatic animals as the bird flies just over the surface of the water. Gulls sometimes have bills that strain the water. The shrike impales its prey on the end of its bill. The night hawk has whiskers and flies with its mouth open to catch insects. Pelicans use their pouch to carry fish to the nestlings, rather than swallowing the fish and later disgorging it into the babies' mouths.

Anatomy has been affected by eating habits in many ways. The vulture who eats carrion has a bald head so its feathers will not be soiled. The secretary bird has strong feet and long legs because it feeds on snakes after it has kicked them to death. A bird called the jacaranna developed long toes that permit it to walk on water-plants while seeking its prey.

It is true, but odd, to say that the songs of birds help prevent overpopulation. After all, song is part of the courtship ritual. Yet it serves other purposes as well, one of which is to indicate a bird's territory, the area which belongs to him. The bird sings to show he is there and will permit no male intruders of his own species. He will defend his territory with fights, usually harmless but among the towhees blood is often drawn. The territory is the amount of room needed to nest the young and find food to feed them. Among eagles a territory may be a number of square miles since they must find larger prey, such as rabbits. Seed- and plant-eaters do not need so much room but, in every case, territories insure that no more birds are born than can be fed. Naturally, birds did not consciously plan matters this way but that is its effect.

Sea birds, who often nest on remote, rocky islands, acquire safety from predators that way and access to rich fishing waters. Their territory is no larger than the distance necessary between two nests so brooding females cannot peck one another. Since they do not feed on land and the sea is so abundant, there is no natural control over population and one small island may have millions of birds on it. This does no harm unless man discovers the sanctuary, as he did a small island off Newfoundland where the fat, tasty auk bred in great hordes, and caused their extinction by killing them for food. Many sea birds do not actually build nests on their territory but the Adelie penguins attempt to because they lay their eggs on slopes which often have snow on the crest which may melt. The cold water running down would kill the eggs so the little penguins build little circles in which to lay their eggs, and these act as dikes to keep the water out. The only material they have to build with are small stones of lava. A male, courting, presents a female with one of these stones as a sign of his interest. Since the stones are in short supply, the Adelies constantly steal them from one another's nest, so the male must continually be on his guard. Penguins, like many other birds, return to the same territory year after year. Clearly they have good memories.

The voices of sea birds (a human can hardly call them songs) are very noisy, perhaps to overcome the sounds of the waves breaking on shore. Yet why do parrots and parakeets let out such raucous screams? To overcome the muffling effect of all the vegetation? Why do roosters crow in the morning and so many kinds of birds chatter just at sundown? Birds are born with the songs instinctive to their species but as they grow they learn more melodies. They often sing for reasons that can have nothing to do with territory or mating. It may simply be that they enjoy singing.

And why do some birds imitate? Margaret Millar, the author, tells of a mynah bird that would keep saying "You're a stool pigeon, Mother." This writer knew a West African gray parrot who grew up in Ghana with a human family that owned a Renault car. Later, after living many years in the United States, the parrot would still make the sound of the Renault automobile horn.

A claim is made that the lyre bird of Australia is the finest singer in the world. A lady who lived near the city of Melbourne named one lyre bird James. They are arena birds, relatives of the bower bird, and his arena was near her house. James liked to entertain and gave one performance for a human audience that lasted 43 minutes, without the repetition of a song. In addition to his own basic song, he sang the melodies of at least 20 other species and also imitated the songs of animals such as dogs and cats. James sang all the year round except during the month when he moulted.

Lyre birds sometimes join each other in duets, in unison or harmony, and an imitation done by one bird is often picked up and repeated by another. The female lyre bird, like her cousin, the bower bird, raises her young all by herself, but she mates for life and her nest is not far from his arena. Certainly she is able to hear him singing and perhaps is consoled in her lonely existence. The songs of birds no doubt carry many social messages we cannot understand. The best we can do is enjoy them.

In the courtship of birds, as in most animals, the female is likely to be coy. When she approaches, the male is apt to be threatening at first, until his usual instinctive reaction is replaced by another. The female coyness is partly because the threats are expected but also because she is exercising her right to choose the best male and will not be truly receptive until she is convinced. Wild birds are seldom, if ever, raped. Once the signals have been recognized, the male often makes an offer of nesting material. He will also go through all sorts of exaggerated gestures, puffings, swellings, singing, flying about, dancing in rituals that differ in particulars between every species. The male who fails to perform the courtship rites correctly will not arouse the proper hormonal response in the female and in this case no mating will occur. If all has been done properly and the male is sufficiently gorgeous, according to his species, the mating will proceed and all will be over in a few seconds. The number of eggs ultimately laid is usually, but not always, the same. A snowy owl lays more or less eggs depending on the lemming population in her northern territory. Since it can hardly be a matter of will power, the egg count must depend on the quality of the female's diet.

While it is usually true that male birds are larger than females, the opposite is the case among predators such as hawks and falcons. Darwin wrote that females evolved as larger in these species "for the sake of overcoming other females and obtaining possession of males" and it is true that predatory females are fearsome. Another suggestion has been made, however, that these females are bigger because the task of feeding their young is such a great one that they need the extra strength. Yet the males do help with the feeding.

So conditioned are humans to certain patterns that the relation between female and male phalarope seems very strange. Not only is the female larger and more brightly colored but she is extremely bold. These are water birds and here she whirls herself in front of him. He timidly turns away. She shrieks furiously, flies against him and shoves him repeatedly with her bill, then makes short flying passes at him as if to attack again. If he accepts her, then all is well, but if he does not she will continue to push him around until he submits. Then he meekly follows her to shore, where they mate. When the nest is built, it is the male who sits on the eggs and feeds the young until they can fly. The entire job is his. In this species the female does the singing.

How did the traditional sex roles change? Speculation is that among ancestral phalaropes both species had bright plumage that was hormone dependent. Some mutation created females that had no urge to brood and incubate. The species would not die out, however, if males took over and there must have been males with enough female instinct in them to feel the necessary urge. These females with a new hormone balance would be free to leave the nest and thus less exposed to predators. More would survive to reproduce than would normal females. The most inconspicuous of the males on the nests would be the ones most likely to live to reproduce. The theory of evolution sometimes has very neat and tidy answers.

The psychology of birds has been the subject of considerable investigation. Obviously they are capable of learning. A blue jay needs only one experience of eating the monarch butterfly, a very toxic insect, to know the creature on sight and avoid it ever after. Learning is also the basis of social hierarchy in birds, the peck order. This is most familiar to us in chickens and probably results from overcrowding. A domineering hen pecks all the others and is never pecked in return. The hen next in rank submits to abuse by the most dominant but she herself abuses all hens below her in social position. A leader in the flock may have a favorite who has no prestige but is given privileges because of this peculiar friendship. The peck order, curiously enough, does not depend on size or intelligence but is apparently a matter of who is most spirited. Among chickens in a flock, the males tend to ignore the dominant females and prefer those who belong to the bottom of the social ladder. Among jackdaws a female of the lowest class suddenly becomes a leader if she has been chosen by a dominant male. Social orders such as this, based on fear, may save us from glamorizing and sentimentalizing animal life. They are not better or worse than human beings. Some animal behavior seems very comparable to that we know among people; other things seem quite incomprehensible.

Among birds there is the broken wing trick, which compares to nothing in man's activity. Birds who nest on the ground, frightened by what seems to be an enemy, often flutter away, pretending to be hurt, and luring the danger away from their young. One opinion today is that this is displacement behavior, not gallantry so much as confusion between the urge to flee and the desire to stay and protect the infants. Another curiosity among birds is the matter of imprinting, which Konrad Lorenz has written about extensively. In the course of his research several of his subjects have fallen in love with him. Apparently he was the first creature who impinged upon their consciousness and so they identified themselves with him. Lorenz had a pet jackdaw that always followed him about. As an adult, this bird tried to feed him caterpillars and when Lorenz would not take them in his mouth, the jackdaw tried to stuff them in his ears. Imprinting is not universal in birds, however, since a cuckoo growing up in an oriole's nest does not try to mate with orioles later but manages to recognize his own cuckoo kind.

The bird mind is primarily an emotional one. Obsessive as a defender of territory, as a suitor, and as a parent, birds are also very capable of jealousy, vanity, and ego, as well as play. Though we can never completely understand them, they will no doubt continue to fascinate us.

When the ripe and fertilized eggs are ready, the female is seized by a new mood. She becomes subject to a new plan or pattern of behavior. Sometimes this may be no more than an urge to fly around in the dusk broadcasting the eggs into the damp herbage. . . Or the egg-laying female may become strongly drawn to some odor that is characteristic of the surroundings which their larvae will require; so that blowflies, which earlier were visiting flowers and feeding on nectar, are now allured by the smell of carrion; and the females of the Chalcid wasps, parasitic on the larvae of blowflies, are likewise attracted to the smell of carrion where their hosts are likely to be found . . .

The Sphegid wasps prepare a live but paralysed insect for their larvae to feed upon—and so on in increasing complexity and endless variety. We see once more the convergence and concentration of the whole life of the insect upon this final act in the reproductive process. It is here that the most elaborate forms of insect behavior are to be found.

—V. B. Wigglesworth, The Life of Insects.

15

*H*ow can one forgive evolution for producing mosquitos? Or cockroaches? Of what use on earth are cobras or even rattlesnakes? In human terms, the earth is full of such riddles. At times it does seem as if there are many creatures on earth who are of no use to man at all. Should we get rid of them?

Can one possibly love nature when it produces females who devour their mates? European scorpions pass a night of love together in some dark place but in the morning only the female emerges. She has devoured all of the male excepting his shell and his body will now nourish the eggs he has fertilized. The male praying mantis approaches the female very cautiously and if she moves to attack, he flees, but if she remains quiet he continues to approach and then in one leap jumps on her back. If he is awkward and does not fall into exactly the right position, the female grabs him about the neck with a front leg and proceeds to eat him, head first of all. Yet the male struggles very little. His efforts are to get his abdomen in position to consummate the act. In this he usually succeeds, though at the cost of his life. Those males who have succeeded and then escaped will take a chance later with another mate. Among spiders the female may also kill during mating. One may well wonder what evolutionary function such practices perform and, even

167

more, how such things came about in the first place? Yet the species do survive and even prosper.

And the insects have certainly learned how to survive. Altogether about one million species have been named and several thousand more are added to the list every year. Estimates are that about three million species of insect actually exist and the number of individuals is in the thousands of billions. So many species exist because they have filled more different ecological niches than any other living thing. Their success is due in part to the fantastic variety of mechanisms for offense and defense.

One common defense for insects is concealment. We see only a fraction of the actual population. The 17 year locust spends all but a few weeks of its long life buried a foot or more in the ground. Some insects protect themselves by hiding deep within fur or feathers. Others such as lice have very tough skin that is hard to crush. The house flies save themselves by their very fast reaction time. (Who has not been frustrated when trying to swat an alert fly?) Caterpillars often act stiff and dead when threatened and frequently look like small twigs. Camouflage is a very customary defense and sometimes a variety of species all develop the same kind of camouflage. There are several kinds of larvae, moths, and caterpillars all of which look just like pine needles. There are butterflies and grasshoppers which exactly mimic certain leaves. The fidelity of the detail in some of these copies is so exact that people sometimes wonder if natural selection could have been responsible. Many insects have developed cryptic patterns with various colors, stripes, and so on that make them difficult to distinguish against the backgrounds they choose. Some caterpillars change their marking as they change their locations; the same animal can resemble a birch twig or an oak twig, and if the oak is covered with lichens, it can look like that too.

Camouflaged insects have also developed bright red or yellow hind wings that are usually concealed but may be used if the animal is discovered by a bird. These wings have spots that look like pairs of eyes and the insect uses them to startle an attacker.

Insects are not merely passive. The mandibles, the claws, the blood sucking parts by which various insects get their food can also be used as weapons. Many insects are venomous. Presumably their various poisons appeared by chance in the animal's chemistry and, proving useful, the new substance was kept and improved upon. Many wasps have poisons that paralyze their victims, usually spiders or caterpillars, and the wasps then lay their eggs on, in or near the helpless creatures, who will stay alive but unable to move until the eggs hatch and the young wasps begin to feast on them.

The mosquito-bite itch does not come from a poison but from saliva that is injected into the wound. (Mosquitos attack not only humans but birds and animals.) This saliva contains foreign proteins to which the skin reacts, becoming more sensitive with each bite. The venom from a bee bite, used only for defense, contains a chemical that sets off a reaction in the human body that produces histamine. It is the histamine shock in persons sensitized to bee bites that can sometimes be fatal. Many ants and other insects also produce poisons that, while not always lethal, are usually quite irritating or painful. Venomous insects often do not bother with protective coloration. On the contrary they are very conspicuous, as a warning. The black-and-yellow banded caterpillar is avoided by birds; and mealworm larvae, which are not distasteful to birds, have adopted this same color scheme and so are also avoided. Birds generally choose to eat insects which are dull in color.

When a species has become too successful, too abundant because it is unpalatable, some other species will come along that is able to modify its eating habits to take advantage of the situation. Thus the European cuckoo has learned to specialize in caterpillars with poisonous hairs that other birds will having nothing to do with. No species is ever completely free from danger. It is just that some characteristics give a better chance of survival, and so they become established.

The insects of the world and its plants live in a state of perpetual warfare. Probably all plant species are attacked by insects and these are so numerous, so varied and specialized, so hungry that it is a wonder plants have been able to survive. The point is that plants have created their own defenses. Cacti have sharp spines, holly has sharp leaves, poison ivy is a poison, and the odors of spice plants are repellent to most predators.

Plant defenses include a wide range of biochemical compounds. Before he learned to synthesize such things himself, man had discovered that pyrethrin powder, from chrysanthemums, was a powerful insecticide and was also harmless to mammals.

Of considerable interest to man are the alkaloids produced by some flowering plants. Among these are nicotine, caffeine, quinine, opium, marijuana, and peyote. The last three all create hallucinations and one may wonder about the behavior of an insect on dope. Eating any plant-alkaloids would have a serious effect on the insect physiology.

Some of the most rapacious of the insects are the 15,000 species of butterflies who, in their long caterpillar stage, have enormous appetites. Caterpillars are very specialized in their eating habits and their choice of food is a matter of the plant's chemistry. Thus there is a cabbage butterfly that will eat anything which produces mustard oil, such as Brussel sprouts, radishes, and watercress. Most caterpillars never feed on any of the 10,000 species belonging to the coffee family. Presumably they have no taste for alkaloids.

Yet some caterpillars do live off plants that produce alkaloids. As butterflies, these insects are brightly colored and birds avoid them as being distasteful. Such species have found a niche with little competition, and they have taken the plant's repellent substance and incorporated it defensively into their body. Many insects that feed on toxic plants are more or less immune to man-made pesticides. It seems they have evolved methods of detoxifying all sorts of foreign chemicals.

This complex, chemical battle between plants and insects accounts for the immense diversity of species in both. They appeared together in the Cretaceous period and evolved together. As the flowering plants developed chemical defenses, the insects diversified to overcome this. In his own long battle with insects, man might intervene in plant evolution and breed varieties that produced chemical repellents, a better method of protecting them than the use of dangerous insecticides.

For human beings, the fact that insects exist has serious consequences. They are not all bad. Ants have a function in growing plants since they aerate the soil. Bees make honey, wax, and pollinate flowers and crops. Social insects (bees, ants, termites, and wasps) are philosophically instructive, examples of successful societies based entirely on instinct but, for this reason, societies in which there can be no progress. Of all the insects, man has cultivated silkworms for the longest time; and the silk industry was once very profitable.

Yet these are not the most important matters to man. What counts is that insects feed upon the plants he grows and therefore they are pests. Getting rid of them by DDT and other poisons, which seemed so promising, has turned out to have many drawbacks. Not the least is that DDT seriously interferes with bird reproduction, and birds are among the most effective controllers of insects. Of course, other methods of insect control are being tried, such as the production and release of millions of sterile male screw worm flies (which very effectively reduced the population) but insects will certainly always remain a problem to farmers and gardeners. And then there are the plagues such as the

171

gypsy moth, boll weevil, and the swarms of locusts that sometimes bedevil Africa and the Near East. The temptation is to use very drastic methods to get rid of insects; and yet all the biological relationships in the world are so complex, the results of actions so hard to predict, that caution must be used. While much of the world is near starvation, and malaria is still a major disease, many nations do not want to ban the use of DDT, and this is perfectly understandable. Yet a consequence of the widespread use of DDT is the evolution of species of mosquitos and other insects resistant to it. Man is never going to rid the world of insects completely; and it might be a terrible mistake to do so even if he could. We know too little about the web of life, the balance of nature, to consider such an experiment.

This discussion of creatures undesirable to man began with the question whether or not we should get rid of them. This may be silly but perhaps it is not. Who will defend poisonous snakes, or man-eating sharks, two animals that have terrified men for many ages? Fear of snakes is so ancient that it is instinctive in monkeys. Even Raymond Ditmars, one of the world's great authorities on snakes, could write a whole book about them without saying one kind word except that they were interesting.

In human terms perhaps the only defense that could be made for snakes is that they eat rodents and thus keep the population down. No one knows how important this is; but, in the absence of sure knowledge, how can one tell.

It has been said that predators have an absolute value because they keep natural selection in operation. They eliminate the weak and defective, thus insuring that only the finest of the species preyed upon will survive to reproduce. This principle has certainly operated all through the long history of nature and should not be lightly shrugged off. It is perhaps the only principle that could be used in favor of the shark. Although they are not the only dangerous animal man can find in the sea, sharks are certainly the most hated. Perhaps it is their size, their emotionless savagery, their swiftness, or maybe it is because they attack in large packs. The early evolution of sharks produced animals so suited to their life that they have evolved hardly at all for many millions of years. Supposedly they have little intelligence, but they do have deadly instinct. Perhaps it is a bit silly, too, and abstract to talk about getting rid of them all but, if it were possible, I believe many ocean fishermen and swimmers on Australian beaches would be in favor of it. When discussing whether life has a purpose, the question of the purpose of sharks does not seem too academic.

It seemed that animals always behave in a manner showing the rightness of the philosophy entertained by the man who observes them . . . Throughout the reign of Queen Victoria all apes were virtuous monogamists, but during the dissolute twenties their morals underwent a serious deterioration.

—Bertrand Russell, My Philosophical Development.

16

*B*etween animals and mankind the relationships are infinitely various. At times as we observe them it seems they must be feeling the same emotions we find natural to ourselves. Most obvious of comparable emotions, or at least behavior, is that between some of the adult mammals and their young. Not only wolves but grown porpoises and baboons assume part of a community responsibility for everyone's children. Most mammals are very social animals, a trait quite understandable to humans. All need some education, not only in hunting but in conducting their emotional lives. Infant monkeys deprived of their mothers' care become frightened, irritable adults incapable of getting along with other monkeys and usually failures when it comes to reproducing. Orphaned infant humans raised in nurseries, with only perfunctory emotional attention, grow up with the same misfortune.

Love evidently exists among many mating couples in the world of mammals. Lions and lionesses display tenderness that goes far beyond the mere act of sex itself. The union of male and female elephants takes days to complete, with every evidence of tenderness and regard. The male elephant has little to do with rearing the young, but he does remain with the herd and act as guard for the entire social group. When it is time for her to give birth, the female elephant is watched over by several other females of the herd. The infant becomes a herd responsibility if anything should happen to its mother.

Dogs, of course, are very emotional about people but are seldom observed showing great affection for each other. Perhaps this is because, like Chi-Chi, the panda in London, they have gone through the strange psychological phenomenon of imprinting and do not really know they are dogs. Cats, supposedly so aloof, are also capable of tenderness and the author once had a female tabby, well domesticated, who most certainly fell in love with a stray tom she met in her basement. The tom would roam and be gone away for weeks and, in his absence, she would go down every day to look for him, sit waiting by the door, and often cry, though she was not in heat at the time. Why should anyone suppose that the higher mammals have no strong emotions. After all, such things did not spring full blown in humans and had to develop from some previous mammal experience.

Social taboos have made human beings not acknowledge a kind of animal behavior that, healthily, they possess. Of course, the training to become an adult human is so involved and takes such a long time that the parents do not always drive away their young after training but actually, when the young human begins to mature, it also starts to compete with the adult and a tense situation can exist. Among mammals, the adult recognizes that it is time for the young to leave the nest and the decision is sternly effected. The subordinate social position of the young is intolerable, and ordinarily they are willing to go. Interestingly, this performs the evolutionary function of making the subordinates pioneer in new habitats. Here they may be able to express themselves and, if conditions of isolation should arise, make a beginning toward new species.

Man has obviously used animals for all sorts of purposes. Someone, who knows when, even had the idea of training lions, tigers, and bears to perform for entertainment. But for all our intimacy with them over the ages, some aspects of their behavior remain mysterious. Why do the adult males of many large carnivorous species kill and even devour their young? Members of the bear, the cat, and dog family all do this at times. Among bears, the female almost always is wary of the male until the cubs have gained a certain size. One explanation is that the males somehow realize that the territory is overcrowded, that other controls have not worked, and that it is necessary to become cannibalistic in order for the species to survive. Male animals caged in zoos are particularly prone to this behavior. It may be that the unnatural surroundings trigger the same instinctive reaction —but none of this is subject to scientific proof.

Overcrowding among laboratory mice and rats has produced similar unnatural behavior, and all the normal mating and parental behavior breaks down into chaos. One of the most famous aberrations among animals has to do with the lemmings of Scandinavia and Alaska. These usually calm little animals are extraordinarily fertile and if they have a few years with good food supply and no diseases, they overbreed until their range is extremely overcrowded. Then the lemmings become very neurotic, fight among themselves, and at last begin to rush off in migrations resembling panic. They run heedless of each other and many are crushed in the race, which is so blind that they often continue straight into the sea, where they are drowned. The human race has been observed doing things not too unlike the suicide of the lemmings; and the motives, and the mass hysteria, may not be too different.

When considering the relationship between man and animals the most obvious fact is that we are all living together. When the relationship is a close one, the ornate word "symbiosis" is sometimes used. When the relationship is of benefit to both participants, it is called "mutualism" and a very simple example of mutualism is the lichen which is actually two different organisms living together.

177

Part of the lichen is a fungus and it provides the living structure in which an algae abides and produces food for itself and the fungus by means of photosynthesis. Other mutually beneficial relationships are those between the cleaning fish and the fish he serves by removing their parasites and other growths. Varieties of pecking birds do the same kind of thing, usually with hoofed animals. (To be exotic one might mention the red-billed oxpecker of South Africa who survives by such tasks as removing the lice and ticks from impalas.)

Other symbiotic relationships go by the name of "commensalism" which means "eating at the same table." Here the benefits are all one-sided. An orchid living on the branch of a tree does nothing for the tree but benefits by having a place to live. Cockroaches do not help people but people unwillingly provide cockroaches with warm homes and food.

When the partnership causes harm to one of the species, the matter is called parasitism. Dogs suffer from parasitic fleas. Man suffers from the parasite malaria bug. The malaria organisms cannot live without the man's blood but the man can do without the parasite. Parasites may kill their hosts, but not immediately. When the killing is done quickly it is called predation and man has often been called the greatest predator of all.

Thus man practices three different kinds of relationships with the living environment around him. He gives his dog food, shelter, and affection and in return has a hunter, a watchdog, or a friend. He carefully cultivates wheat, corn, and cattle but in the end kills them. He consciously shoots ducks, geese, and deer and in the past slew mammoths.

All of this has gone on since life began and is nothing to become sentimental about. (One sympathizes with vegetarians who abhor killing but even they must eat plants that have been killed or they would not long survive.) Why should anyone be sentimental about killing anything then, since it is part of nature itself?

What man has added to natural relationships is his conscious and unconscious power to kill in immense quantities. A lion kills only one zebra at a time. During the nineteenth-century hunts for the passenger pigeon, James Audubon told of a man who claimed to have killed 10,000 of the birds, mostly with nets, in a single day. Even if the man was boasting his kill was of a different order of magnitude than that of any other predator. And of course such hunts for the passenger pigeon led to their extinction.

But what of extinctions? What does it matter that we have no passenger pigeons? After all, most of the species that ever lived are now extinct. Two-thirds of the more than 2500 families of animals that once existed have died out without leaving any descendants at all. Yet the average species had a span of about 70 million years before they disappeared. Humans have wiped out more than 150 species in the last fifty years. And 14 species of mammals, 46 species of birds, 8 kinds of reptiles, and 21 different species of fish are now on the official Endangered Wildlife List of the United States Department of Interior.

Some extinctions have been deliberate, as when the white man set out to destroy the buffaloes of the plains to deny them to the Indians. Others have come about through less organized but nevertheless persistent hunting, but hunters, at the same time, have been partially responsible for the conservation of some game birds and fish through the enactment of laws limiting the number that may be taken. Man has brought about the present crisis in his relation with other animals not just by hunting but by what he has done to the environment they must live in.

As a geological agent, man has changed much of the face of the world. By building a canal around Niagara Falls he built a water passage not only for ships but for the primitive sea lamprey, which proceeded to enter the Great Lakes and nearly destroy all the native whitefish and lake trout. By creating the Aswan dam on the Nile, the eastern end

179

of the Mediterranean that was once refreshed by river water is now becoming increasingly salty and warmer so that its fish population is being seriously reduced. Dams have increased the hazards of life for salmon on North America's West Coast and now there is talk about a sea-level canal between the Pacific and Atlantic, through central America, allowing sea life from the two oceans to mingle. This could create a man-made ecological disaster of dimensions impossible to calculate. Geologically, men have left great holes in the earth from open-pit mines, stripped the Appalachian hills bare while mining coal, covered over the land with vast cities and immense strips of concrete road. And now oil companies propose to lay a pipeline to carry hot oil in Alaska over the ground with its shallow permafrost. Not only would this hinder the migration of caribou but the heat could melt the ground and thus create landslides, fracture the pipeline, and cause oil spills over the landscape. To fight a war, Americans have scraped, poisoned, and pockmarked huge areas of Vietnam to keep the land from the enemy; and no one knows how many centuries it will take to cover over the scars.

The story of man as a polluter is so universal and has been told so often that now it threatens to bore people. But it does not resemble the story of the boy who cried "Wolf." When water is actually listed as a fire hazard, as is the case of the Cuyahoga River in Cleveland and the Houston ship canal, then pollution has become a clear and present danger to everything that lives. A random selection of newspaper and magazine articles collected within a few months gives a picture of the problem's dimensions:

Special Danger Seen In Arctic Pollution
[From *Marine Pollution Bulletin*.]
Sea Dumping of Waste Off San Francisco Banned
Aroused Europeans Try to Stem Industrial Pollution
Is The Mediterranean Dying?

The heavy hand of man has snarled the fragile web of
life in the forests, moors and heathered
slopes of Scotland
Lake George Fish Unsafe to Eat Because of Mercury,
State Says
U. S. Lists 91 Beaches As Closed or Polluted
Biological Nitrogen Fixation in Lake Erie
Scientists Caution on Changes In Climate as
Result of Pollution
DDT Traces Found in Antarctica
[It has never been used there.]
Fish in Sea Losing Fins To Sludge
Dutch Scientists Blame Polluted, Dredged-Up Mud for
Widespread Deaths of Seabirds 100 miles Away
Petroleum Lumps On The Surface Of The Sea
Shortage Of Caviar
[Pollution of the Caspian Sea.]
British Naturalist Says U.S. Is World's Biggest Polluter.
"Each American accounts for more toxic wastes poured
into rivers and oceans than 1000 Asians."
Denmark Reports Big Decline In Storks,
Her National Bird
1000 Acres of Smog-Afflicted Pines To Be Cut
[Lake Arrowhead, Calif.]
Sea Slicks Contain 10,000 Times More Pesticide
Than Surrounding Water
Astronaut Says Earth Is in Need of Protection
from Inhabitants
Senators Hear Mercury Is Peril to Fish Industry
Akron Detergent Battle Seen as Vital
The Global Circulation Of Atmospheric Pollutants
Galapagos Verging on Desecration
The Pines of Ravenna. "A doomed and haunted
forest outside an ancient city."
DDT Residues in Marine Phytoplankton.
Increase from 1955 to 1969

181

The End of Civilization Feared by Biochemist
"A Harvard biochemist says civilization will end
within 15 to 30 years unless immediate action is
taking against problems facing mankind.
Dr. George Wald, a Nobel Prize winner
made the statement. The 'overwhelmingly threatening
problems' are pollution, overpopulation and
the possibility of nuclear war."
Tidal Wave of Babies Is Forecast By An Expert
"The country is now experiencing our second
postwar baby boom, an echo of the first."

Such stories are depressingly familiar. That some of the
concerned people speak of pollution and population at the
same time is obviously because each new human is another
producer of wastes that must be somehow disposed of.

If so many people are so troubled by the problems of
pollution, population, and the wildlife that is vanishing
(really all aspects of the same question) why is nothing done
about it. Much is being done, of course, but it is not at all
clear whether mankind will win the race or whether it will
manage to defeat itself.

Besides, many people do not seem to believe much of any
problem exists. This comes as a shock to members of the
fashionable "ecology" movement but is unquestionably a
fact. The Mayor of Fairbanks, Alaska, said it would be
"anti-God" not to take out the oil that God put in the
ground, no matter what the ecological cost. The fight is
really a very old one and is between people with very different views about the grand design of life itself.

Consciously or unconsciously, many people do not care
or bother to think about the consequences of their acts.
Such people interfere with nature as if it did not matter
in the least. Biological results of such attitudes can be quite
astonishing. Someone decided that Australia needed rabbits,
imported a few, and they prospered so well without serious

enemies that they became a plague. Someone else imported dandelions into the United States because they thought they were pretty. How many million gardeners have paid the price of this folly? In 1890 a bird lover and admirer of Shakespeare thought that the United States should have every bird in the country that the great poet mentioned in his works. Some did not succeed here, but a flock of starlings let loose in New York City thrived beyond belief in the new environment. The gypsy moth is not native to North America. It was introduced here in 1869 from Europe by a naturalist in Massachusetts who wanted to study them. Now, with DDT on the way out, every tree in New England is threatened by gypsy moths and they have even reached Texas. The walking catfish, relative of the African lungfish, suddenly appeared in Florida and Georgia in 1968. It can move across land from one body of water to another by use of its large pectoral fins. At first it was thought that it attacked dogs and people. This turned out not to be true, but it is a strong predator that goes after shrimp, snails, tadpoles, and smaller fish. Attempts to eradicate it have proved futile because it can simply walk away from trouble. The walking catfish, say Florida authorities, is here to stay. In 1970 California authorities were shocked to discover the fish in one of their lakes. Apparently imported from Asia by pet-shop owners, the fish presumably just walked away from a dealer's pond.

Florida seems to take the brunt of many alien invasions. Water hyacinths brought from South America for their decorative value reached the waterways of the state and now have spread toward the Gulf states, clogging rivers and canals with their lush growth. In 1966 a boy returning to Miami from a trip to Hawaii brought his grandmother three giant African snails as a present. By 1969 twenty thousand of these snails infested a 13-block Miami area, decimating the foliage and even eating the paint on houses. In 1970 a boy fishing in the Florida Everglades caught a red-bellied piranha, the vicious little animal that belongs in South American rivers. Presumably this piranha was also an escapee from a pet shop. Officials hope it was an isolated case.

Heedless about throwing nature out of balance, many people even fought to save DDT when it was proved beyond doubt that the insecticide was destroying species of birds and fish. To some, animal deaths have no meaning. A famous African elephant named Ahmed who lived in a forest in Kenya for many years, and even starred in a film, had to be protected by a special order from the President of that country after he learned that two American hunters were planning to make a special trip to Nairobi just to bag Ahmed in order to get his two famous tusks. The rhinoceros in India is still being slaughtered by poachers, in spite of official protection, because its horn can be sold for high prices to Chinese who believe that, when powdered, the horn acts as an aphrodisiac.

Former Secretary of the Interior Hickel banned the import of any products made from the eight severely endangered species of whale. The whale was used mainly for cat food and as an oil base for soap, beauty creams, and such. The day after Hickel was dismissed, the ban was lifted. No explanation was given. In any case, Russia and Japan continue to hunt whales without regard to the future.

To the conservation minded, the heedless destruction of wild animals cannot be justified merely because someone may call it "sport" or make money from it. Poachers threaten the legally protected alligator of Florida because women will pay well for alligator leather. The snow leopard of Asia will be gone in ten years due to the demand for its fur to be used in coats. Kangaroos are on the way out in Australia. They are killed to make dog food, for their skins, and because sheep farmers call them pests. The wild horses of the West have gone from a population of two million at the turn of the century to 16,000 today. They are also killed for dog food and for "sport." The vicunas of Bolivia and Peru are vanishing, in spite of protection, because Europeans and Americans will pay high prices for cloth made from their wool.

Due to the acts of man, the animal world is upset in many, many different ways. Overfishing caused the end of the prosperous sardine industry in California and is now doing the same thing in the Bay of Biscay off France. Oil drilling and recent home building in a suburb of San Francisco has caused an invasion of rattlesnakes. One husband and wife between them killed 39 rattlesnakes around their yard in a two-month period. The wild elephants of Ceylon are down to a population of no more than 2500 because their forest and jungle habitat is rapidly being cleared for agriculture to feed the fast growing population. During food shortages in China in recent years, Mao Tse-tung ordered the killing of all the dogs, cats, and birds. Then rats, mice, and insects devastated the crops. The cat population of Colombia in South America nearly vanished as the result of indiscriminate use of pesticides and fumigation chemicals. This led to an enormous increase in the rodent population, many cases of rat bites and incidence of the diseases carried by rodents. An ordinary house cat in Bogota now sells for $12, a week's wage there for a construction worker. The balance of nature, the ecology, is very complex and no one understands it well; yet people recklessly continue to tamper with it.

The story of man and nature is not entirely negative, fortunately. The pipeline in Alaska has been halted for further study. Three species of animal endangered in their native lands, the ibex of Iran, the Barbary sheep, and the African oryx, have been brought to New Mexico and released in regions "too rough" for other animals; thus they will upset no existing balance and they all are thriving. Many other rare species are being preserved in parks but in impoverished countries they are endangered by the human population. Rangers try somewhat hopelessly to guard the game in Africa while in India the only 150 specimens of the

Asiatic lion still in existence find the starving "untouch-ables" driving them from their kill and eating it themselves. Lion cubs are starving and the population dwindling,

In 1970 New York State passed a law forbidding the sale of a number of furs including leopard, tiger, cheetah, vicuna, polar bear, margay, and ocelot and also the sale of alligator or crocodile skins. The harvesting of a 10,000-acre stand of timber was halted in 1971 for a year because it appeared that the supposedly extinct ivory-billed woodpecker had been heard in the woods. Golden eagles hatched in England in 1971 for the first time in 100 years, and the heron appeared on the Thames in London. It had not been seen for many years and presumably returned because the river's pollution was being somewhat controlled. Perhaps just in the nick of time, estuaries such as Chesapeake and San Francisco Bays, where so many birds and fish breed, are being studied to protect them from polluters and land fillers. A number of refuges have recently been established to help save the prairie dogs of the American West. The United States Army was prevented from killing millions of blackbirds, that were annoying many farmers and some civilians who worked at an ammunition plant in Tennessee, by freezing them. Many farmers found so many birds a nuisance to their cattle and pigs and favored the extermination but others did not and publicized the planned event, causing national pressure to stop the Army and the Department of Interior which was cooperating in the extermination. Alaska has curbed the hunting of grizzly and polar bears. Pakistan has set aside a 190,000-acre area near the Ganges to try and save the beautiful Bengal tiger.

But preserving nature or manipulating it is seldom easy; the main reason being that one also has to deal with people. An example of this is Greenland, which is consistently getting colder. The Eskimo in this Danish colony lived on codfish and seals but the codfish are retreating south before the cold and the seals are virtually extinct. To save the

186

Greenland population of 50,000 people the Danes introduced reindeer and the animals thrived there. Yet the Eskimos are hunters and refused to herd them, conserve, and build up the stock. To save the Eskimos, the government is thinking of bringing them to Denmark, though they realize this would create a minority problem familiar in other countries.

The complexity of manipulating nature shows up in the problems of the Murchison Falls National Park in Uganda. Here "elephant damage" has set up a controversy among the conservationists themselves. Elephants turn woodland into grassland by pushing over trees, debarking them, and overgrazing new growth. They endanger a diversified habitat for themselves and other species. One zoologist recommended that 4000 elephants be shot in this park where they are protected, about 40 percent of the population, in order to protect the remainder of them. The same man also wants 6500 hippopotamus along the Nile River killed, to prevent overgrazing. Some Uganda officials reluctantly agree with this proposal, but similar officials in Kenya and Tanzania say the problem can be worked out without resorting to such slaughter.

The Central Park Zoo is getting a replacement for Skandy, the polar bear that was shot and killed on June 3 after grabbing the arm of a man who had climbed a fence and stuck his hand into the cage . . .

The empty cage has become a major attraction for visitors since Skandy was killed. Zoo visitors, who had often been charmed by the polar bear's cavorting with an aluminum beer keg in the pool outside its cave, have placed many memorial tributes at the cage. There are fresh and faded flowers, memorial wreaths, poems, posters, sketches—provided by children, men and women who had loved the polar bear. Some of them said the policeman who rescued the man should have shot him, "not the polar bear."

There are conflicting reports about the incident. Some witnesses said that the man seized by the bear had been maltreating the animals and birds at the Central Park Zoo. Others said he had simply been trying to feed them.

—The New York Times. June 20, 1971.

17

The incident of the slain polar bear provokes a number of thoughts. Here is an animal held captive through no fault of his own, then killed when he did what any bear would be expected to do. The guard, a human being, probably could not stand by and see one of his own kind clawed or bitten to death but why was he put in the position of having to commit such an injustice? What was in the mind of the guilty man, the stupid man, who exposed himself to such a danger? Was he teasing the bear? How could he hate it? Did the bear represent some kind of symbolic evil to him or was the man merely showing off, demonstrating he did not fear the animal? Was he contemptuous of the animal because it was behind bars? What is the function of zoos, if at all?

People who are indifferent to the fate of animals are incomprehensible to those who care for them. To Buffalo Bill and his comrade hunters the bison was nothing but a thing to be exterminated. In the town of Viola, Minnesota, each June for the last 97 years they have held a Gopher Count, a party that begins with a parade and ends with the killing of as many gophers, rats, skunks, and crows as possible. There are people who acquire animals to amuse them at their summer home or cottage and then in the fall when the people return home, they simply abandon the summer pets. Some residents of Capistrano in California destroy the nests of the swallows that have made the town famous because the swallows are "a nuisance, a terrible mess." Also in California the fishermen would like to wipe out the sea otters, once before almost extinct because they were hunted for their fur. The otters compete with the fishermen for abalone, the good tasting shellfish, and so it seems to them the beautiful otters must go.

The grizzly bears in Yellowstone Park have been known to attack people, either because hikers unknowingly came too close to a mother bear and her cubs or when a bear wandered into a camping grounds looking for garbage to eat. It is not enough to suggest that humans in the Park be shown how to avoid such incidents, it is proposed that the grizzly bears be exterminated or at least moved somewhere else (although this is the only place now where they roam free.) A correspondent of *Science* magazine, writing about the bear problem and other dangers of Yellowstone, such as the geysers, asked: "After we remove the threat of the grizzly attacks are we going to turn off the hot water?" When the Prince of Wales remarked that the hills of his principality looked very bare without any deer on them, such as he was used to seeing in Scotland, Welshmen rose up in indignation at the thought of their land being stocked with deer. "Do we want to produce food or an open air zoo?"

The American Indian once thought of the animals as his brothers and killed only what was necessary. Animals all had spirits too. Now some writers suggest that such a view of nature was destroyed by Christianity. Taking the Biblical view of man's superiority, one could exploit nature with indifference. The Pilgrims in New England looked on the forests as something to be cleared for farmland and as dangerous because Indians could hide in them. Plants and animals were thought of as existing simply for man's use or enjoyment. Opposing this view that Christianity is all to blame for this attitude, others point out that many people who are not Christians also ruthlessly exploit nature. It is natural for man to be "anthropocentric," to believe that everything revolves around him. His first thought is for himself. Is it possible for him to think in any other way? Can any of us really be saints?

Writing in *Science* magazine (December 4, 1970) Dr. Charles Southwick of Johns Hopkins raised the alarm about declining primate populations and what their scarcity would mean to biomedical research. Apparently 100,000 primates, animals such as the rhesus monkey of India, are used every year in United States medical laboratories alone. The Indian villagers are becoming increasingly intolerant of rhesus monkeys because of the food they take. In their poverty they are eating the rhesus monkeys, which they also use for medical potions. Dr. Southwick is worried about what will happen if the present level of *harvest* continues; evidently they are nothing but crops to him. He wants to avoid the extinction of *any desirable animal* but he does not say who is to set the standard for desirable animals. Who is to decide? Someone who can talk about harvesting them? Having written about the problem of an adequate supply of primates, Dr. Southwick concludes by warning against any undue emotionalism about primate conservation. When the shortage of primates becomes generally known, he is worried for fear that biomedical research will be blamed.

"In 25 years all the wild animals, perhaps three-quarters of all the species living today, will be extinct." This was said by S. Dillon Ripley, Director of the Smithsonian Institution. It caused ripples, for the most part, only among those prepared to hear. Man considers himself so unique that what happens to others simply does not matter. In the pioneer days of the United States the land and every resource in it and living upon it were things to plunder. When things were used up, you could simply move on. Private greed was the expected conduct, and the man who made the most money was the most admired. The biggest anything was the best, and this attitude has not changed very much. To a businessman, his sacred duty is to produce dividends and if this means polluting rivers so no fish can live in them, that is simply tough luck.

Measures are taken to control pollution and smoke and pesticides, but they are not very effective since no one wants to pay the cost. Citizens vote against sewage plants because they will raise their taxes. Factories pay small fines for smoking chimneys and continue to commit the same offense. Farmers want pesticides because they want good cash crops.

To many people the saving of nature is merely effete and a sentimentality. Leaders in the conservation movement around the turn of the century were mostly rich men, and they were suspected of trying to rope off the wilderness to keep others from making money on it and to preserve it mostly as their own private luxury. Conservationists were thought of as an elite group intent on getting an intellectual satisfaction. Of course the national parks are enjoyed by millions of citizens now, to the point of serious overuse, but they mean nothing to the poor in cities who cannot afford to go to them. Preserving nature does not seem very urgent if you live in the slums of the East Bronx in New York; and the cries of those concerned about it are greeted there with apathy or suspicion. It would be nice to swim in New York's rivers but not if taxes have to be raised in order to make it possible. The good people of Montana love their hunting and fishing and clean water but the residents of the town of Lincoln are boycotting the general store of Cecil Garland and trying to drive him away because he led a drive to prevent a huge mine from being opened. He thought that the mining development would be an ecological disaster. It could have meant 800 to 1000 jobs. "It sure would have been a wonderful thing for the valley. That payroll would have helped keep us from going broke."

Is there anything in nature worth saving if some man thinks he needs it? If the answer is affirmative, how do you prove it to the man? Can it be shown to a man that there is any purpose in life other than his own satisfaction? Can we find a grand design other than those offered by the great religions? Does the story of evolving life and the constantly changing face of the world have any point for human beings or should we merely ravish everything as we desire since no point exists except our own pleasure? Is there any reason to save the Bengal tiger from extinction?

We know what the tiger evolved from but does that explain his purpose? The tiger's history and ours began together, in single cells of organic matter, capable of dividing into two, that came either from outer space or were generated in a primordial broth in a shallow sea somewhere here on earth. With this beginning, the general principles of what we call evolution immediately began to go into effect.

Not all the beginning cells were equal. Many living cells died only moments after creation, but those that were more fit survived. They grew larger and divided into two. Then, as part of the mystery, two cells learned to combine and form a new being. The magic of sex had been discovered and, with this, the possibilities of development became theoretically infinite. Since the two combining cells were not exactly alike, the result of their combination would resemble one cell in some ways and the other cell in different aspects. These cells were subject to the stresses and strains of the radioactive, chemical, hot and cold atmosphere around them. And thus, at times, they changed, underwent mutations. Most mutations were harmful but some made the living things better able to cope with their surroundings.

As time went by the reproducing organisms began to take on considerable variety and to compete with one another for food and living space. And always there was this inescapable urge to reproduce. The individual always struggled to preserve its own life but the purpose of this was to survive and generate more of its own kind. Often, in nature, the individual dies immediately after it has performed its reproductive act.

As evolution continued, each species became more and more complicated. This was a competitive response to changes in other species and those species which could not keep up with the pace all perished. The world kept changing its form, sometimes isolating on islands species which had been in the mainstream of the struggle. Safe from competitors, island species sometimes changed in ways that rendered them defenseless and thus, if challenged later by those who had developed aggressively to survive, the island species were quickly eliminated.

Finally, after 100 million experiments with species, man with his self-conscious brain emerged. He believes that he is the most complex of creatures and superior to all others. The fact that he has such thoughts means that he probably *is* superior, in many ways. No bear, no lion, no elephant can successfully compete with man as a species and, where man wants to live, other animals must leave or submit to man's domination. What kind of laws brought this about?

The laws of evolution are not absolute, like those of physics. They are historical. Everything that has happened before is the cause of whatever happens now. All history is contemporary history. There is no way that anything can ever be turned back to what it was in the past. If a species develops a specialty, takes a certain road, it must live with it. A horse cannot have hands because it has hooves. A bee can never read a book because its species has elaborated a life that is completely instinctive. Man has succeeded, comparatively, because his major specialization, his brain, does not block his destiny but makes him extraordinarily adaptable instead. Unspecialized man is the most omnivorous of primates and the only one capable of surviving in such diverse places as the South Pole, the Equator, seven miles down at sea, and on the surface of the moon.

The competition of life is not only between species but

within them. The strongest walrus or elk, the most beautiful pheasant or peacock, these are the most likely to win a mate and pass their genes on to the following generation. This kind of selection may be even more important than that between animals of different kinds.

Darwin's theory, in part, was the result of his reading Thomas Malthus's essay on population. Malthus's dire opinion was that poverty in the world is inevitable because population increases by geometrical ratio and the means of subsistence by arithmetical ratio. The only ways population growth is halted is by war, disease, famine, or moral restraint. (Only humans practice moral restraint, birth control, but the custom is hardly universal.) As Darwin applied Malthus to the natural world, he saw that competition between the species was the check that also prevented overpopulation.

The study of evolution shows a trend toward filling every available niche. Nature abhors a vacuum. A new species may attract a parasite that also eventually becomes a new species by adapting itself to its host. A new volcano, such as Surtsey in the sea off Iceland, soon becomes another field, sown by airborne seeds, where plants begin to grow.

A prominent function of life is death. Without it there would be no room for more involved, higher forms of life. On a larger scale, extinctions of species serve the same purpose. Extinctions come about for many reasons and these are not always readily evident in the fossil record. Recently the idea has been advanced that extinctions come about because of changes in the earth's magnetic field. Periodically in the past the North Magnetic Pole has reversed itself, a change that takes a few thousand years. During the ebb, the earth's magnetic field weakens and, briefly, even disappears. During this time the earth is exposed to lethal rays from the sun that are normally excluded by the shield. Studies have shown that various species of radiolaria, microscopic little animals in the sea, have become extinct at the same time the magnetic field was at its weakest. It is suggested that birds or fish or any animal that responds to magnetism would be disoriented, too, and this might lead to their destruction. These ideas are not yet accepted by all scientists but they do point up the interdependence of things in nature.

Part of the story of life is the story of the environment in which it is lived. Long before man began to alter the landscape, plants and animals had been doing the same thing on a much slower scale. Much of the surface of the earth is made up of calcium carbonate which is nothing more than the remains of once-living animals. The soil which supports us is made up in large part of the remains of organic material. Animals burrow and dig holes, beavers build dams, elephants tear down forests, goats have devastated much of the lands bordering the Mediterranean. All these changes and many more like them cause new changes in their turn.

It can be readily seen that nature is neither gentle nor kind. But it is not cruel, either. Emotion only began to exist with the coming of birds and mammals. Even with these creatures we can only guess at the depth of their emotion. Many acts in nature which are cruel in human terms have no emotional content in the animal's life. Does a lion feel guilty for killing a fat zebra or an osprey for catching a fish? Who cries when presented with a perfectly cooked rare steak?

One of the proofs of evolution is the incredible memory system within every living thing. A rose knows how to be a rose, a porpoise how to swim, a redwood grows several hundred feet in the air, all without conscious effort. Man breathes, digests food, reaches puberty, labor pains begin; we sweat, have goose pimples, recoil from sudden danger, all without thinking about it. Some such things are common to many species; others are quite particular to one. All of them are products of natural selection; mutations that were preserved, long before there was anything known as deliberate thought, because they had utility, for the individual and thus for the species, in their effort to keep alive,

A search for the design in life reveals that nothing is ever fixed or permanent. Constant change is the rule. There is no conservative status quo. Everything is in the process of becoming something else. Throughout history, life has tended to expand, to include increasingly greater numbers of living things. At times life has contracted, and all sorts of species become extinct, but the pressure is for more, more, more.

In this eternal battle, a way to survive is to find new ways of making a living. The early mammals were small and resembled the unattractive shrew. After the dinosaurs had gone and the arena cleared, their descendants radiated into many different zones. Mammals took to the sea, others commuted between land and water, bats took to the air, other mammals changed in order to eat the newly evolved grasses, and the ancestors of man took to the trees. Some of these mammals ate nothing but insects, others leaves, some ate other mammals, and some ate everything. In accordance with evolution's rule that size is an an advantage, the most successful animals were the ones who grew larger.

Evolution has opposing trends and one is toward specialization. Some animals became so particular about their diet that they became fatally dependent upon it. The koala bear cannot survive on anything but eucalyptus leaves and, therefore, in a world dominated by human beings, the koala would have a very slender chance for survival if he was not gentle, helpless, and therefore appealing. Being fit for survival takes many guises. Not until modern man, however, was being cute one of them.

And now, effectively, natural evolution may have come to an end.

The human race considers that it has won the battle. In addition to their often-mentioned adaptability, human beings have the highest reproductive efficiency of any animal. They have successfully replaced all competitors. All the many experiments with the very superior chimpanzees show that there is an enormous chasm between their understanding and ours. In terms of future evolutions, it is certain that human beings, as long as they thrive, will never allow any other animals to cross this chasm. If any species seriously challenged man's supremacy, that species would be quickly and very efficiently wiped clean off the earth. That is why man is probably evolution's end.

Of course mutations will continue to happen frequently, but for the most part they are very small. At present they usually go undetected but any serious changes will be noted and taken care of. When the day of biological engineering arrives, mutations will be controlled by man. (Not everyone agrees that this will be an unmixed blessing.) Toward what direction will he aim his own evolution? How can we know man's proper goal in nature?

Some years ago the author asked Albert Einstein if he would be willing to contribute his thoughts to a magazine article, then in preparation, called "Why Was I Born?" In reply, Dr. Einstein wrote, "The question 'why' in the sense of 'to what purpose' has, in my opinion, meaning only in the domain of human activities. In this sense the life of a person has meaning if it enriches the lives of other people materially, intellectually and [or] morally." Einstein did not reveal how he arrived at his opinion.

The question asked in the title of the book is whether a purpose can be found in the story of evolving life. Among the clear trends of this evolution are all the adaptations for survival and for an ever more complex brain. How does a complex brain help in survival? Don't insects who live entirely by instinct survive equally well? Maybe a conscious mind has some other value than mere survival of the species.

Only man could have the idea that there might be a grand design. Only man does design, only man can be a conscious artist. This ability came about through evolution but how are we to understand it? Why has man become the deliberate creator of things?

One of his obvious creations is the society in which he lives. His rational faculty makes it possible to record and transmit his culture, a second system of heredity. He can pass on to later generations all he has learned; his many skills, his life-saving medicine, all his pleasure-giving arts, music, painting, cooking, and so on. Man is the only creature who knows he has a history and he transmits this record too. Human history is full of horrible events, wars, murders, folly, but if one studies the record there does seem to be progress. Material progress is obvious but there are those who believe that democratic systems of government represent progress over tyranny and there are those who believe they can see improvement in the relations between races and between nations. We no longer accept misery as inevitable. Infectious diseases are now under control. The emotionally ill can be successfully treated rather than locked up in asylums. We do not now fear natural disasters, such as hurricanes as the wrath of God, but can predict them and protect ourselves to a degree.

Julian Huxley, the great biologist, believed that evolution had at least a direction, a line of progress, one that should guide us in planning our future. These trends were general ones. Increase of control. Increase of independence. Increase of internal coordination. Increase of knowledge, of the means of coordinating knowledge, of elaborateness and intensity of feeling. Huxley wrote, "If we do not continue them, we cannot hope we are in the main line of evolutionary progress any more than a sea urchin or a tapeworm."

Most evolutionary progress has led to blind alleys in which future development is impossible. The human brain presents a quite different outlook. With it experience can be handed down. Rather than be directed by the accidents of evolution, random mutations, man can now control it.

A Frenchman wrote, "Mankind, alone in the universe, is not finished." Man can speed up changes in nature at will. He can make evolution conscious. *Evolution produced a creature capable of giving it meaning.* Evolution gave the future to man and, if his endowment is used well, he has a right to be optimistic.

Evolution has not given us a morality but it has made us capable of having one. It did not give us a God, even though we may yearn for one. It almost denies the existence of a God. Theodore Dobzhansky, the noted American geneticist, wrote, "If evolution was designed and is managed by a supernatural force, it is managed remarkably badly." Man sees that everything has a reason and, in seeking a reason for existence, it was perfectly natural to create a God and to create Him in man's image. How else could man imagine that God might be? And children in western civilizations grow up believing that God looks like the man in the painting Michelangelo made on the ceiling of the Sistine Chapel. Those who believe must do so on faith.

One thing clear about evolution is that nature constantly strives for more complex organisms, but to what end? Conscious intelligence seems preferred over blind instinct and all those animals nearest to us have it so some extent. The conscious intelligence is not necessary for survival. Why should nature keep working toward it? Perhaps a more complete awareness, not mere survival is what nature was aiming at. Awareness, to a much higher degree than any other animal, is what man has. Since it allows us to control, it must be a goal of nature. And, being aware, we know what pleasure is and are able to create it. A further thought is that awareness makes us conscious of love, able to feel love. Animals, too, feel love to some extent but we believe that human love is more intense. A human loves more than just one master—he may love his mate, his children, his parents, and his friends, all a bit differently but all at the same time. He may also love abstractions such as his religion or his country. Love is the emotion that is impossible to define—but that hardly denies its existence.

Evolution has led us to an awareness by which we are able to conceive of good and evil. It does not give us our moral code because ethics themselves have had to evolve out of our *experience*. Wolves have a moral code but salamanders do not. Good and evil did not exist in the time of the dinosaurs. All tribes, all races believe that incest is evil because they have seen the bad results in the next generation. Various foods are taboo in different religious faiths very often for the good reason, such as pork among Jews and Mohammedans, that it keeps badly and may make people sick. The moralities of the Ten Commandments make good practical sense, in part, because the rules help keep the peace between neighbors. "Thou shall not covet thy neighbor's wife" is very sound advice.

Evolution does not provide man with a goal but it has given him the means to find a goal for himself. Thousands of years before Darwin, man was already directing biological and social evolution. He has been playing God for a long time. Up till now, however, the role has been played without much of a script and without much respect for the laws of nature. But while man may have become dominant in the scheme of things, he is still a part of nature and not yet its absolute master. There is a complicated balance between all living things he knows so little about, that he is always flirting with disaster as he alters the balance. There is also a relationship between man and earth itself and here man has methodically extracted and despoiled as if there would be no future generations. The search for a grand design teaches us that man is still very dependent on nature.

Our moral, unscientific, and historical conditioning cries out that it cannot be part of the design of evolution that after three billion years it finally brings forth a creature to destroy every wild living thing about him! Our social evolution demands that this not be true. The design thus far has produced a creature capable of knowing, controlling, able to create his own pleasures, and able to love. That is obviously part of what it has been about. Evolution has devised a creature capable of creating his own morality, and part of this demands that we preserve nature. Common sense agrees. And this same evolution has developed a creature who can design the future to his liking. That is also the achievement.

But in reaching his present level, man developed traits that could eliminate man himself. The species is not immortal. No guarantee exists that the victory will be permanent. Human minds cannot really imagine the extinction of their species. It is not one of our capabilities to really imagine even our own, individual deaths. But even if we cannot conceive of extinction, many of us can see how life could become unbearable long before such a point was reached. The words like pollution, overpopulation, depletion of natural resources, vanishing species, are all too familiar.

The day has passed when every man can have the freedom of Daniel Boone. People have accepted many rules in order to live together; rules about noise, about litter, about the use of their cars, about a hundred different matters. But some rules for ths social good are very difficult to enforce. The United States went through a very sad experience when it tried to prohibit the consumption of alcohol. How much more difficult it would be to try to limit the number of babies a man and woman may have. Yet some day, perhaps, people may accept this as necessary, although the force of public opinion may cause people to restrict their families voluntarily before the world becomes practically uninhabitable.

What worries many people more than overpopulation is the degradation of nature itself. Most sad about this is the condition of the animal world. If you stop polluting a river, in time it will clean itself up, but when you have killed the last black panther, they are gone forever. The scientific way to look at animal life is that it is a gene pool, "a vast reservoir of adaptations to countless environments, physical hazards, and diseases" and this is tremendous insurance for the future. The majority of those concerned about animals are less cold-blooded about the crisis. They feel a sense of identity with animals simply because they are alive and, as the Indians believed, they are our brothers. Children are taught to love pets and neither hurt nor tease them and in time the love for an animal can become a very strong emotion. By transference this feeling extends to all animals. This is not just weak sentimentality but something very powerful. As we mature and our aesthetic sense develops we understand the beauty of a thoroughbred horse, the grace of a tiger, the joy of a sea lion, and the wonder of undersea life around a coral reef. We believe such things matter in their own right. They are their own justification for being and we want to save them. Many species will certainly be saved. Those particularly endangered in impoverished countries where animals are either a nuisance, or looked upon as food, can be moved elsewhere. Men can be their custodians, until the day arrives when they are appreciated in their native lands and returned to live there, protected and as sources of national pride. (The Greeks permitted Lord Elgin to steal many marble treasures in the last century but now they want them returned.)

Those animals that man finds useful will be protected and survive. Many of the rest will be gone soon. Even if they mean nothing to us in terms of beauty or pleasure, how can we know they will not be useful in some unseen way in the future? It will then be too late. Man may now be in effective control of evolution but he can never reproduce what it took three billion years to create.

Perhaps the economic and biological common sense of preserving other species will be understood by enough people so that most of them may have a chance. Those to whom it is an emotional matter will certainly keep up the pressure and already the emotional ones have been able to command enough eloquence and power to make their views strongly felt.

But there is another matter concerning evolution, now becoming very much a part of contemporary news, that has a far greater emotional content than even the present drive to save the wild animals of the world. This matter is the growing awareness that new techniques and new under-standing of genetics may ultimately destroy sexual repro-duction in the traditional sense.

Through direct manipulation of individual genes, in-herited diseases might be gradually eliminated, and human evolution speeded toward the creation of a race of super-men. Through other techniques, direct physical contact for reproduction could be eliminated (already a fact in artificial insemination) and a world of test-tube babies, such as Aldous Huxley's *Brave New World* terrified the reading public with in 1932, would become the norm. The emerging knowledge called genetic medicine presents wild possibilities for man's future. The emotional, social, religious, biological, and psy-chological consequences of *planning* new generations can be either horrifying or inspiring. Actually the new power now rapidly emerging by which man can take even more direct control over his own evolution may be the greatest challenge the species has ever faced.

At present the most obvious intervention in reproduction is artificial insemination. It is routine in the breeding of horses and cattle. In humans it is done when the prospective father is infertile. No accurate figures exist but many thou-sands of children in the United States have been born in this way. The anonymous donor is selected for his normality, his physical resemblance to the infertile male and for some-thing called his "superiority." Obviously this practice has produced legal complications and, one may well imagine, distress in the ego of the man who has been found incapable of reproducing himself.

Recently a new development has taken place in the field of artificial insemination. Some years ago it was found that sperm can be frozen and the first human conceived this way is now seventeen years old, a boy in fine health who is an excellent student. In 1970 over 5 million cattle in the United States were produced with frozen sperm. Now, beginning in the Twin Cities, commercial establishments have been started that will store a donor's sperm in a frozen sperm bank, for a fee, until he calls upon it. Many men who are about to undergo vasectomies, surgical sterilization as a means of birth control, first make their sperm available in case they later decide that they do want to have children. Others donate to the bank in case an anonymous donor of their type should be wanted at a later date. The most unusual donor, perhaps, was a man in Minneapolis who had his sperm frozen on the chance that his one son might turn out to be sterile and might want to use his father's sperm to impregnate his own wife, to keep up the family's line.

When sexual selection operates freely in a state of nature, the most fit members of that species have the most reproductive success. This operates in humans as well; the strongest, the finest looking, the most aggressive, will be the most likely to have children. But humans have recognized generally, and scientists confirmed in particular, that there are many defects hidden in the prospective mate that could prove disastrous in the next generation. Since human beings as a species have more than 1000 harmful recessive genes that could be damaging if the partner had the same recessive gene, efforts have long been made to avoid such unions. As mentioned previously, there are many clinics to help people worried about the problem but a new method, not just historical and physical, promises more exact findings on the question of whether a child will be born with genetic problems.

This is the somewhat drastic method of drawing off amniotic fluid from the womb of a pregnant woman and analyzing it for some of the more common conditions. If the foetus is found to have a negative Rh factor, this may be corrected by injection of an intra-uterine fluid. If an enzyme deficiency known as PKU is discovered, preparations can be made to give the infant a corrective diet immediately on birth. Should the foetus turn out to be mongoloid, an abortion is recommended—if conscience and the law permit.

In the future there may be a procedure called artificial innovulation to discover hereditary defects. The idea is that a foetus in a very early stage of development could be taken from the mother, without damage to either, kept briefly in a test tube while a few cells were examined, then returned without harm once more and re-implanted in the mother's womb. If the foetus had too serious defects, it would simply not be re-implanted. A further idea about innovulation would be that the foetus might be implanted in a donor mother who would nourish it and keep it alive until the rest of the nine months were up. This extraordinary idea would be for women whose health was not good enough to bear the child herself. Or perhaps for women who were unwilling to go through the tedium of pregnancy. (Most proposals in genetic medicine shock our traditional values of what is natural and right. Women, however, have breast fed other womens' babies for many centuries.)

Although the idea is at least 50 years old, the idea of babies nurtured entirely in a test tube still seems radical yet increasingly possible. Not long ago researchers at Columbia University kept a human embryo alive for six days in a test tube and a scientist in Italy kept one alive for two months. The latter began to develop into a monster and was destroyed. After pressure from the Vatican, such experiments have ceased in Italy. A scientist, commenting on test-tube babies, gave the opinion that they would be psychological monsters, even if physically whole. This would be because they lacked the security of a womb; but development is now under way to create an artificial womb that would duplicate the human womb in every way.

With artificial wombs, intercourse and the role of the male could be eliminated entirely. Using a method called cloning, from the Greek word for thronging, thousands of cells, from an admired individual, could be grown in test tubes, then transferred to the artificial wombs for the remainder of the growth. The result would be thousands of identical twins. Asexual reproduction is a very technical matter, hard to understand, but it is possible. With it, by cloning, the world could have as many Heifetzes or Mohammed Alis or Thomas Edisons or even Stalins as might be desired. Such spectacular evolution is no longer science fiction. Whether or not it is desirable is a matter that should be seriously debated before we discover that scientists somewhere have secretly and, as they did with the atom bomb, actually gone ahead and worked out the techniques.

In the few years since the structures of the genetic messengers (DNA and RNA) were worked out and artificial genes produced in the laboratory, the possibility of genetic surgery has opened up. It should not be long before harmful hereditary genes can be detected before pregnancy has begun. When these can be spotted, it may be possible to remove them by some yet undiscovered method and replace them with healthy, externally supplied genes. Obviously there are mechanical problems involved and whatever procedure was developed would take much of the passion out of reproduction; but it might be worth it to avoid such hereditary problems as diabetes, hemophilia, or a strong tendency to schizophrenia or cancer. People may be willing to put up with a good deal to eliminate the roulette quality that is part of having a baby today.

Less obviously desirable is what genetic surgery could do, not just to eliminate hereditary defects, but to alter the species for the future. Once genes can be altered to get rid of something undesirable, they can just as well be altered to add something.

Without intending to be funny, it has been suggested that it would be worthwhile for man to have two stomachs, so he could digest cellulose, thus utilizing a new, abundant source of food for the ever-increasing population. It has also been thought desirable that man should have a larger head, so that it could hold a larger brain. It has also been considered that the species might be deliberately evolved so that babies would be born with the innate capacity to recite the multiplication tables, as birds today are born instinctively able to navigate by the stars.

More modest than such proposals, we might begin by eliminating the remnants of our evolutionary history that are no longer useful; get rid of the adenoids and appendix, for instance. We might also change the genes to produce stronger backs and to avoid the production of wisdom teeth.

At least two cautions need to be made about these wonders-to-be. The production of thousands of identical-genius-twins might homogenize the population to a dangerous extent. Genetic variability would be lost and genes eliminated, that while no use at all under present conditions, might be highly desirable in coping with a changed environment. As in the example of the recent corn blight, if all the species is identical, it will all be susceptible to a mutant disease to which no individual had the germ of immunity.

The other caution for people who expect immediate miracles from the new biology is that some conditions are caused by a single gene while others derive from a combination of many genes. Abnormalities such as hemophilia are caused by one gene and will thus be comparatively simple to eliminate. Qualities such as intelligence, however, are the product not only of numerous genes but are affected, as well, by the individual's physical condition and the social environment he finds himself in after birth.

Speculation about what man may do with his own evolution is by no means a parlor game. The process of changing natural means of reproduction is well under way and much research is going on that will bring even greater changes. Among animals and plants under man's direct control, "natural reproduction" can hardly be said to exist at all. This majestic control contains the power to do immense harm. But wise men also have the power to create a world much more desirable for all species than any world that has ever been known before.

More than anything else, the grand design worked out in three billion years now provides a starting point from which the future can at last become rational. The grand design need no longer be a blind force, with rules, but no end in view. Human beings have already taken the grand design into their own hands.

As a species, man is paddling through rapids more perilous than the wildest river. We know there is broad, calm water at the far end of the canyon. Is there enough wisdom in our species to navigate the course?

I must believe that there is.

Index